Restoring Our American Dream
The Best Investment

Michael K. Farr
with Edward B. Claflin

Foreword by Brent Scowcroft
Afterword by P. J. O'Rouke

Headline Books, Inc.
Terra Alta, WV

Restoring Our American Dream: The Best Investment

Michael K. Farr with Edward B. Claflin

copyright ©2013 Michael K. Farr

All rights reserved. No part of this publication may be reproduced or transmitted in any other form or for any means, digital, electronic or mechanical, including photocopy, recording or any information storage and referral system now known or to be invented, without written permission in writing from the publisher, except by reviewer who wished to quote brief passages in connections with a review written for inclusion in broadcast or print media.

To order additional copies of the book, or to contact the author:

Headline Books, Inc.
P. O. Box 52
Terra Alta, WV 26764

www.HeadlineBooks.com
www.MichaelKFarr.com
www.OurAmericanDream.com

ISBN-13: 978-0-938467-66-3

Cover photo by Philip Bermingham

Library of Congress Control Number: 2012955619

    Farr, Michael
    Restoring Our American Dream
      p. cm.
      ISBN 978-0-938467-66-3
    Non-Fiction

PRINTED IN THE UNITED STATES OF AMERICA

For Laurie, Robert, and Maggie, I love you up to the sky!

For the United States of America, a great, shining city upon a hill and for all of my fellow Americans. Our greatest and best days are still ahead of us!

And in memory of my wonderful father-in-law, Robert Nelson Fishburn. Bob was an extremely bright, gentle, loving soul who adored his family, was devoted to his friends, and was passionate in his philanthropy. Thank you, Bob. Your life made a difference. We love and miss you dearly.

CONTENTS

Foreword: A Yearning for Dignity by the
Honorable Brent Scowcroft 5

Introduction 8

Chapter 1—A Fair Chance at the Jackpot 13

Chapter 2—How We Conduct Ourselves 25

Chapter 3—The Psychology of Reward and Punishment 48

Chapter 4—A Level Playing Field 60

Chapter 5—When Trust Erodes 75

Chapter 6—The High Price of Corruption 94

Chapter 7—Side Effects of Hope 105

Chapter 8—Responsibility for Consequences 118

Chapter 9—Earning Trust 134

Chapter 10—Going for the Dream 148

Afterword: by P. J. O'Rourke 165

End Notes 168

Acknowledgments 175

Foreword
A Yearning for Dignity

by the Honorable Brent Scowcroft

I would not be offering to write the foreword to this book if I did not share with Michael Farr a significant concern. Indeed, that concern is the very topic of *Restoring the American Dream*, and it is my hope that after reading what Michael has to say, you will be prepared—in your own way, by the means that you consider most effective—to do something about it.

We are talking about honor. It's one of those antique concepts that, I realize, is generally absent from public consciousness. To me, however, it is the most important of a triumvirate of words that helped guide my path at the educational institution I attended as a young man. I refer, of course, to West Point, where the motto engraved on granite portals and in the mind of every cadet is "Duty, Honor, Country."

It's remarkable that the mission, strategy, goals, tactics, ideals, and vision could be embodied in so simple a motto. For an institution like West Point, with its reputation for turning out leaders who have taken America through some of its toughest and most challenging times, wouldn't you expect an elaborate set of regulations and instructions? Shouldn't there be a set of laws and bylaws with extensive instructions on what does, and does not, constitute duty; on what is meant by honor; and how one should behave in service to one's country? Instead, there are just three words embodying the principles guiding an institution that has ably served this nation for several centuries.

At certain points in America's history, this simple but challenging code has been embraced not just by West Pointers but by a great number of U.S. citizens. Certainly World War II was one of those occasions when every man and woman serving in the armed forces, every family on the home front, and every institution and organization in the republic lent their support, energy, determination, and resources to the single-minded defense of our country. From those years of shared sacrifice and unity of purpose, there emerged not only millions of unrecorded examples of individual courage but also the ascendance of leaders like George C. Marshall and Dwight D. Eisenhower who had the ability to carry the nation forward from a terrible wartime into an era of peace that posed its own formidable challenges.

Today, that code is upheld with great sacrifice and courage by those who voluntarily put their lives at risk in service to our country. Yet, despite the wars we are fighting and the many challenges we face on the home front, there is no single great cause uniting our nation. I agree with Michael Farr that this is a time when America is being severely tested, and I think the American people can meet the test, but we need to reassert the hopeful fundamentals of what this country is all about.

I am a believer in America's exceptionalism. We are a unique country because we're not ethnically or religiously pure. We're exceptional because we have pursued, much of the time, what I would call "homely virtues"—that you shouldn't expect anything that doesn't come by hard work, that you can have a better life than your parents did if you work and learn and improve yourself. That's what we have always been taught in this country, in contrast to many other parts of the world where you expect to live just like your parents did and expect your children to live just as you did. In America, we have our place in life, but we also have the notion that we can all do better. We have hope and a yearning for democracy, but even more so—like people all around the world—we have a yearning for dignity. People want to be treated like human beings, with respect. They need to know their rights are not being trampled on, their portion of the national wealth is not being stolen. The American way of life, I believe, should always offer a way to achieve the dignity that we long for.

But ours was never a government designed to make all of us prosperous. It was designed to *allow* you to become prosperous if you worked hard within the rules. And what were the rules? They forced you—if you will—to be honest and not cut corners. In a sense, those rules reflected what is so well summed up in the West Point Cadet Honor Code, that a cadet will not lie, cheat, or steal, or tolerate those who do.

In recent times, we have become increasingly suspicious about sectors of our society and government where lying, cheating, and stealing have become everyday occurrences. This change is well described in this book. What arises, in this atmosphere of suspicion, is a question of trust.

How can trust be restored? I believe we must set an example. I recall the phrase that was used by our Founding Fathers (and repeated centuries later both by John F. Kennedy and Ronald Reagan) referring to the ideal of "A Shining City on the Hill." Ours was to be an example to the world of the ability of human beings to govern themselves. Today we have gotten into the habit of telling other nations "You ought to be like us" or "We'll tell you what to do." We have arrived at a place where we analyze human rights in all the other countries of the world. We compare other nations to ours, and we conclude that in many respects we are better than people who don't live in the kind of atmosphere that we enjoy in this country. But we do not analyze ourselves on the same scale. And we are still far from achieving what we have been striving for—the example of a "Shining City on the Hill."

What will it take to get on track?

However complex the issues, in order to restore trust and hope, it is necessary for us to agree on fundamental principles. Michael outlines some of them here, and I urge you to read what he proposes. As for my own recommendations, based on my experience, I've found that the simplest principles are often the most reliable. Out of personal habit, and my respect for a great tradition, I continue to advocate the three that I adopted when I passed through the granite doorways of West Point at the age of eighteen. If we can keep them in mind—and act on them—perhaps that's all the architecture we need to resume building that "Shining City on a Hill."

hopelessness. They didn't feel like they had a fair shot at opportunity. They felt like the world was fixed and they were unjustly screwed. I was horrified at the hopelessness of these young folks. Additionally, they were intent on voting not for those who promise to safeguard possibility and opportunity, but for those who would provide for their needs.

John Winthrop, the first governor of Massachusetts, implored the colonists to make their new community a "city upon the hill" in the most Biblical sense—a place that could not be hidden, that would be the light of the world. As Brent Scowcroft noted in his foreword, this phrase has been picked up and repeated by many American leaders. But a shining city on a hill is more dream than reality. So what are we really looking for? What is the dream that we, as Americans, always seem to pursue?

For some, the dream needs to be spelled out in economic terms: it's the opportunity for power and prosperity. For others, it involves basic human rights—realizing what lies at the root of the promise spelled out in the Declaration of Independence that "all men are created equal, that they are endowed by their Creator with certain inalienable rights." But in the feisty spirit of American debate, even these aspirations have been subject to perpetual reinterpretation. (What is the meaning of opportunity and prosperity? Why should inalienable rights be limited to men? Do we have to believe in a Creator in order to exercise these rights?)

So What is the American Dream Today?

In its broadest outlines, it's the hope for what's possible—and for the conditions that will allow you to succeed. It's what you can do when you believe that you can follow your heart and the appetites of your soul.

What we've seen in America, over the past several hundred years, as each generation has reached out to realize their dream, is an extraordinary combination of talent and tenacity—a talent and tenacity that overcomes odds and obstacles and achieves things that were heretofore thought to be impossible. Reaching for that dream, and achieving it, is woven into the stories of millions of Americans who overcame unbelievable adversity and emerged from the ashes—often in a single generation—to realize achievements that no one thought possible. We have not been entitled to have things provided for us. But we do feel entitled to possibility, to opportunity—to have a level playing field that will allow us to seize for ourselves almost anything our minds and hearts can imagine (and our fingers and backs can bear).

America really is a place of the dream and imagination—where heart, soul, and hard work come together to create miracles. It is the heroism of being the first college graduate in your family or inventing a phone that plays music, takes pictures, and surfs the internet.

If we tend to think and speak of this Dream with spiritual fervor, so be it. There is a kind of American religion that is woven into our creed as a nation and our spirit. This attempt to create a Valhalla and Eden is of our own doing and within our grasp. It won't be limited by governments, dictators or religious extremists. Fundamentally, we share a spiritual belief that we should have the rights and liberties that are necessary to pursue happiness and achieve it.

Of course we also have the flag, songs, symbols, currency, military force, and all the political and economic clout of nationalism. But none of these accoutrements speak to the real spirit of the American dream. That spirit is in your story, and mine—the story of one person, or a family, who came to this country early or late. They came for many reasons and by many avenues—of their own free will, under the duress of economic hardship, in flight from persecution, or under the yoke of slavery. Having arrived here, all began over again with whatever goods, skills, hopes, or intelligence they brought with them. They carried the spark of human dignity and the will to survive.

Whatever family or personal story you have to tell is yours alone. I'm sure it is filled with tragedy and grief as well as success and happiness. But if you live in America, the story of you and your minuscule shares of hope, are the employed and unemployed, the office worker and the laborer, the full-time wage-earner and the part-time employee. There is a cheerfulness about them, combined with a dogged patience, a complete absence of surliness or resentment that should be noted. After all, no one other than themselves has dictated they should stand in these queues. Their orderliness is not maintained by police ordinances or military repression. Whether they stay the course and reach the cash register, or look for other opportunities, is entirely their own choice. And even if they are not winners in this great lottery—which about 175,999,999 won't be—they will not be rioting in protest when results are announced.

Each has an equal chance, however remote. Each individual in line has exactly the same opportunity as the person ahead and the one behind. The cheerfulness of their participation is not a product of their circumstances. Some will go home to sveltely furnished houses, dapper families, and gourmet meals. Others will make-do in apartments with stained walls and battered furniture, dining from what ever's left over in the fridge. There are even some hollow-eyed and haunted souls who will be bedding on a cot in a crowded homeless shelter or doubled-up on a sidewalk grate when the day is done. But the distinctions of origin, class, or circumstances do not affect their chances in this lottery. Each one shares an equal abundance of hope.

Does this strike you as rational behavior?

A pragmatic economist (which may be redundant), looking at this flagrant waste of time and productivity, could easily calculate the millions of man-hours and woman-hours that are "wasted" by this horde of patient members of society who, while in line, are doing nothing more productive than marking time while they groom their dreams. A sociologist would quickly mark the delusional nature of this stampede: It doesn't take much interviewing to confirm that each participant does have a fairly realistic view of their chances while simultaneously harboring a vision of the possibility of winning $640 million. A brave statesman might attempt, courageously, to remind the assembled masses they are participating in a giant gamble—no different, in its own way, from rolling dice or playing the slots. A righteous cleric might bemoan the waste of so many good intentions serving Mammon. And journalists will confront them with the fact that the winner, in all likelihood, will bring down on himself or herself a sea of trouble as friends, relatives and remote acquaintances claw for a share of the windfall—which, judging from precedent, is likely to be dissipated in a span of five years or less.

None of this matters to the individual who holds an opportunity in hand.

So let's be upfront about this. When we're dealing with the topic of hope, we're jumping in, with both feet, to a subject that includes delusion sometimes bordering on insanity. Motives are often perverse. The thrower of the dart hopes to hit the bull's-eye. The doer of good works hopes to get to heaven. The Islamic martyr hopes for an afterlife among virgins. And the very first immigrant who stepped ashore in America hoped for a better life.

Let us not examine too closely the statistical probabilities of any of these hopes coming true. That's as futile as trying to convince a potential Mega Millions winner to step out of line. There is something, in the very nature of hope, that is, per definition, a bit nuts.

What's even more confusing are the many ways that hope, so irrational in its foundations, can be vanquished. Explorers of the human psyche and practitioners of psychology would love to know exactly how this happens, because they deal with it all the time. Right there, in the Diagnostic and Statistical Manual of Mental Disorders (DSM-IV) definition of Depression, is the descriptive phrase "feelings of hopelessness." Exploration for the magic pill or set of pharmaceuticals that would turn off hopelessness has been a long, well-financed pursuit, and while there have been some breakthroughs, the goal remains elusive. At the center of the mystery is the question *why*. Why do some people only have moments of hopelessness while others struggle with those feelings almost constantly their whole lives? Why do some people maintain hope in the face of what appear overwhelming odds while others seem to "give up" at the earliest signs of adversity? Why do some people face daunting circumstances with seemingly bottomless reserves of optimism while others, facing exactly the same conditions (at least in objective reality) are enervated at the prospect of confronting obstacles and overcoming challenges?

In an article in *Psychology Today,* "Developing the Survival Attitude," John Call, PhD writes, "A person's attitude has a lot to do with his or her chances of survival in a disaster. Believe that you can do something to make a difference in the situation. If people believe that life happens to them and they can't do anything about it, they won't do anything at all. Believe in yourself and your ability to take action, and your chances for survival are much higher."

John Call was referring to the individual's chances of survival. But for a society to survive, it needs the leadership of those who can instill and share the kind of belief that he's talking about.

***

In 1990, when my Pan Am flight landed in Moscow, I found myself in a landscape of diminished expectations. As the plane taxied toward the center of the terminal, I gazed at the billowing red flag with a yellow hammer and sickle that cast a flailing shadow back and forth across the fuselage. That flag was as chilling a symbol as my baby-boom generation ever knew. But it hardly prepared me for the gloom that surrounded me as I stepped off the plane.

As a visitor to the Soviet Union, I had come to assist in the development of their first stock exchange. I found myself in a nation that felt subdued and demoralized. Gray buildings matched complexions and outlooks. Seventy years of Communism had imbued generations of Russians with melancholy. The greatest goal, for many, was to go unnoticed. Based on their experience with Soviet bureaucracy, many believed that the rooster that crowed loudest found his way to the cooking pot first.

Among most of the people I dealt with, I felt a complete lack of any kind of striving or aspiration. There was little hope that anything could be done to change the day-to-day workings of a system that encouraged resignation. Incentives were meaningless. To exist was enough. The popular joke was that the government pretends to pay us, and we pretend to work.

But there were exceptions. Igor Konstantinovich Klyuchnikov was one of them. Igor was an economist who worked for the government. He saw the possibility of trade, commerce, and revenues for Russia, St. Petersburg, and his comrades. He worked within the Party, formed international friendships, and created the foundation for a nascent stock exchange.

Never heard of him? He never made the cover of *Time* magazine, much less *Pravda,* and within the crumbling Soviet Union, he was probably regarded as a renegade or worse. He was quiet and determined and never sought glory for himself. Igor Konstantinovich harbored and incubated a spirit and vision of capitalistic survival not only for himself but for his country.

And he helped light a fire. Beneath what appeared to be a barren tableau were the embers of hope, the vision of possibilities, and a longing for something better. The ubiquitous gray fog had not, after all, suffocated every expectation. Igor Konstantinovich Klyuchnikov kindled promise in the core of human spirit. Thanks to Igor, I learned there were many who longed for a direction—a determination to make a better life. All they needed was a path. Before long, I found myself working with three hundred aspiring brokers and traders. They were excited about our mission. They had voracious appetites to learn. As we talked about trading and stocks and futures and options and bonds, the handful of hopefuls began to dream of what might be.

Still open, of course, was the question of how to make it all happen. But I could tell they were ready to make the commitment to a different direction. After seventy years of Communism, the decades when state-directed appropriation of resources had failed to address their needs, they harbored an embryonic desire to have a better tomorrow. Igor—the skinny, determined economist—had a message they wanted to hear. It was extraordinary to see how his influence fair and equal basis. What about the equal distribution of winnings? The howls of protest will be heard far and wide, as those who "invested" in the Jackpot protest the petty payout. Or what if the retailers took it all? Suddenly you will see good solid citizens willing to break store windows and smash cash registers. Or the indefinitely-delayed-announcement scenario? Well, as the contest participants wait and wait...and wait...to see whether the payout will ever come, people will simply stop participating in subsequent lotteries. Game over, due to lack of interest.

What we see, in all these screwed-up scenarios, is the diminishment of hope and the paring down of possibility. The opportunities awaiting those in the queues are corrupted by foul play, self-interest, violation of trust, and opportunism. Sure, those who run the lottery can get away with it. After all, they run the lottery! They can even argue that players who participated were deluding themselves by playing at all. Everyone knew their chances of winning the jackpot were close to zero—so what difference does it make if their chances happened to be *reduced to zero*? Looked at logically, it's splitting hairs.

Okay. But there's something else going on here. Something that has to do with the feelings people share about the society they belong to and the ways in which things operate in that society. For a short while, when the game was being played, the opportunities were equal. As long as the rules were known, the game was fair, the legitimacy of the winning number was verified, respect prevailed. Knowing full well their chances of winning were close to zero, participants accepted those odds and participated enthusiastically *as long as they felt confident everyone was playing by the rules and as long as there was a scintilla of hope they could win.*

Remember *The Music Man* film with Robert Preston? It is a musical comedy about possibility, hope, and love. The story is a feel-good triumph of human will over doubt and the improbable. Ron Howard plays an adorable, wide-eyed little boy with a lisp. Howard sings a song about the arrival of the Wells Fargo Wagon.

(Younger readers can imagine the UPS truck.) As he imagines the various treasures that might be delivered to his town, he sings, "It could be something for someone who is no relation, but could be something special just for me!"

Whether or not you believe there's something just for you in the Wagon, it's great to have hope. But hope by itself is not a strategy for getting what you want. Trust, rules, respect, fairness—these are part of the foundation that make hope possible. Yet, hope is the engine. Without it, Soviets remained in colorless flats with eyes downcast and hungry stomachs. With it, you get to share Ron Howard's secret thrill that, despite all logical odds, rules of physics and gravity, maybe, just maybe, it could be me!

The Power of "What If"

When my sisters and I were little, Dad would suggest that maybe, for our next birthday, one of us might like a pony or a monkey or an elephant or all three! I remember my little sister Stephanie's enormous eyes and grin as she asked, "Really, Daddy? Really?" Dad would say, "Really!" and launch into a very serious discussion of where we would keep the elephant and what it would eat and who would feed it. He wondered aloud whether any of us be brave enough to ride the elephant around the neighborhood for exercise? As he went on, we slowly realized that he was being silly, but I assure you that as each of us closed our eyes on those nights we smiled and thought *what if?*

In economics, business, and politics there is incredible power in the notion of "what if." What if some genius company could make a phone that could talk, play movies, record movies, send mail, find dinner recipes, locate restaurants, and help with schoolwork? What if Johns Hopkins and Dr. Fred Smith and Sibley Hospital produced a drug that cured all forms of cancer? What if man could go to the moon and walk on it? What if your country could be safer, fairer, more prosperous, and enjoy peace and tranquility? The power of "what if" is nuclear. Therefore the people who deny anyone the right to say *"what if"* are among the most despicable and destructive people imaginable.

These are the ones who can undermine it all. They're the assassins of hope.

Assassians of Hope

The professional assassins of hope have had notable careers, well-recorded in the annals of history. Slavery, in its many different forms, seems to have been with us from the beginning of recorded time. In Ancient Egypt, the pharaohs established the habit of buying, selling, and trading of people like cattle. Legions of conquering hordes, led by the likes of Attila, perceived the advantage of depriving conquered people of their lands and livelihoods, slaughtering those who showed excessive zeal in defending themselves or stubbornly maintained their rights to have access to food and fodder. Greeks and Romans occasionally talked a good game—in terms of rights for their citizens—but in those cultures, too, the slave was, by definition, a person who not only had no right to hope, but no reason to, either. And the list goes on. The great hope annihilators include emperors, caliphs, tsars, kings, and conquerors. Among them can be counted the mine owners of yore, classically exercising their option of driving their workers to early graves, and the slave traders of today, dealing in human trafficking.

In the past and in the present, we don't hear much from people whose hopes have been vanquished by assassins. That's because speaking up, being heard, gaining recognition, is in itself a form of self-identification, a proclamation of self and freedom that can only be attained (much less sustained) by those who have hope. Those *without* hope are, by definition, the silent ones.

And yet, as we also know from history, there is ultimately no silencing people who have been oppressed. Eons of slavery may pass, but eventually slaves revolt. For centuries, tsars played their serfs like pawns, but eventually the serfs were heard from. The conquered never seem to muster the long-term acceptance that would have been required to maintain subservience forever. The mute eventually find voices. Hope, it seems, is somehow hard-wired into human circuitry.

Despite centuries of change, we are still dealing with assassins of hope. They belong to a tradition of greed that never really dies. Though they don't have the bloodlines of haughty pharaohs and ruthless tsars, they are sometimes just as manipulative. Certain impulses do not die easily—and among the ever-living traditions of humankind is the custom of taking hope from some in order to tip the scales in favor of others.

Since current-day "hope assassins" don't always dress up in Hun-style leather helmets, imperial-style crowns, or dictator-style military hats, they may be difficult to identify. But they are still with us.

Hope is not destroyed by falling on bad times and facing the consequences. That happens to all of us at some point in our lives. There is nothing wrong with a fair system working against you, there is simply sadness. Wrong is when the system fails one-and-all because of the failed stewardship and malevolent greed of trusted, empowered leaders. Wrong, too, is when we individuals fail to do our part, when we surrender instead of strive, are silent when we could speak, or blame others instead of looking to and relying upon ourselves.

Take the mortgage foreclosure crisis, for instance. Some people, quite willfully and knowingly, got in over their heads. But many others suffered because of the actions of a few powerful people. There is a difference between those who enter into contracts they are unable to keep and those who lose everything because their trusted bank— run by a former governor and chairman of Goldman Sachs— unlawfully uses their money speculating in futures markets. Moreover, when nothing happens to that CEO, and you have no redress, we see the modern example of the basest forms of pillage and plunder.

So who are these new invaders – the advocates of property seizures, the assassins of hope who have displaced millions upon millions of families and created an atmosphere rife with the terror of uncertainty? They are not just the rich and powerful, because there are rich and powerful people who use their money and their influence with great discretion to help create a more level playing field for all. Bill Gates, for instance, is as heroic a figure as the modern world is likely to see. His determination and creativity have enhanced productivity and the quality of life for all humanity, and his philanthropy is beyond measure. Because of his efforts, he has grown rich—hugely, enormously rich. Which might make us jealous. But he has not used his power or his wealth to kill the hopes of others. Quite the opposite.

Power itself is not bad. Power used badly is bad.

But if we're going to make judgments about good versus bad uses of power, that means we have to take a look at how we conduct ourselves in public life.

Chapter 2

How We Conduct Ourselves
The Luxury of Morality

Is it ever okay to tell a lie? Have *you* ever lied? Does that make you a liar? If you were caught in your lie, were you able to regain your credibility? Did you keep your friend, or did you have to find a new group to hang out with?

People who lie to us are scorned. They are rejected and ostracized. On reality shows, they get voted off of the island. Fundamental to human relationships is the desire to rely on each other, to know that if someone announces the location of food and water, it will indeed be there. And when we are lied to, friendships and relationships are damaged or they're ended because we can't trust the other person anymore.

There are different types of lies. Some are relatively meaningless and others are matters of life and death. The opportunities to fudge the truth come along every day. In certain contexts, "You don't look a day older" may be just an exaggeration, words that escape your lips without a whole lot of regard for accuracy. No damage done. Your intentions are good, the words are greeted with a smile and, if you're lucky, the response is some equally well-meaning reply indicating that you, too, appear to be enjoying the springtime of eternal youth.

Now let's move to another level of seriousness in truth telling.

When someone asks, "How's business?" your hyperactive mental computer may quickly scan any number of well-intentioned replies before arriving at what your ego, superego, and synapses decide is a fitting tribute to the way things are actually going. Of course, if business is great, it's a one-to-one match with the truth to say, "Business is great." It's only when business is not so great that the mind begins to flit over a sea of possible replies. You *could* say, "Business is great" even when it's not, but does that amount to...a shading of the truth, or an outright lie? "Business sucks," you might also say, but is that any truer? After all, there is always the possibility that a huge contract could come in, or you may experience a totally unanticipated turnaround that makes your negative assessment of the situation look like the nattering of a hopeless pessimist. So you scan the other alternatives, usually keeping in mind the personality, loyalties, and affiliations of the person making the inquiry while assessing the probable impact of what you're about to say and, at the same time, keeping an eye on the reflective mirror that shows you how you are presenting yourself. Serious and honest: "We know the challenges, but we are addressing them." Detailed and obsessive: "The past quarter was down, but the first few weeks of this quarter have shown an improvement of orders of about 2 percent over this period a year ago." The blamer: "In this economy? You kidding? We've been hit pretty hard." The eternal optimist: "Absolutely nothing that we can't handle." The mournful victim: "If people would just pay their bills, we'd be fine." The gate watcher: "Hard to say at the moment."

Another thing to note in this interpersonal dynamic is that the truth, whatever it is, can shift. Let us suppose, for instance, that the person asking you, "How's business?" has been entertaining the notion of investing in your business, has seen it as a good one, and is hopeful that his perception of your prosperity is, indeed, in line with his expectations. Be assured that a reply of "Business sucks" will immediately put him off. And he will be far less interested in further detailed analysis than he would be if you launched the conversation with the response, "Business is great." Or suppose you're talking with a potential competitor who is going to take from your assessment a byte of information that's needed to help him make a significant strategic decision?

In the first instance, your portrayal of the "truth" might lure or deter an investor who could significantly impact the future prosperity of your business. In the second instance, a slip of the tongue could result in a change in the competitive landscape which, too, would alter the profile of your business moving forward. This is the meaning of a dynamic shift: because of the role of the other person involved in this conversation, p0otential outcomes change even as you're describing realities. There is no such thing as an impartial speaker delivering cold hard facts nor is there an unbiased listener feeding this information into a data processing system. You have your agenda; he has his. The problem is not that truth suffers—the problem is that truth goes through a mercurial change in the process of transmission.

How's that for getting abstract?

Turns out, it's not so abstract at all.

What if you were Chairman of the Federal Reserve Bank of the United States? How would you handle the dissemination of truths that can shift and transform as you tell them? You'd be in a tough spot. Think about this: You notice that consumer purchases are getting a bit weaker, and you know if this continues, manufacturers and retailers who find themselves selling less will begin to cut jobs. You also recognize that the data are mild and may be nothing but a small blip in an otherwise stable situation.

At your press conference, CNBC's Steve Liesman asks if you are concerned about the slightly weaker consumer numbers. Oh boy! Be careful. If you say that you're concerned, you know the front pages of tomorrow's newspapers around the world will read "U.S. Fed Chair Worried Over Collapsing Consumer." Moreover, you know when that happens, markets will begin to adjust inventories, and employers will preemptively reduce headcounts. Diminished purchasing for inventories will lead to excess manufacturing capacity, and those folks start to fire people too. Then, those people who are out of work stop spending and stop buying things, and voila, consumers are indeed slightly weaker. Nice work, Bigmouth!

Let's think about your alternative answer: You tell Liesman that you have no concerns about the U.S. consumer. Tomorrow's headlines will proclaim "U.S. Fed Chair Bullish Over Strengthening Consumer." Manufacturers start to hire and increase production, retailers start to hire and increase inventory. However, because the consumer was indeed weakening, demand doesn't actually measure up. Supply becomes excessive and there are insufficient revenues to support the new employees. So, layoffs ensue and the outcome is the same as above, but this time you've forfeited a good chunk of your credibility.

This Fed Chairman gig is clearly fraught with peril, and the consequences can be global. Can you imagine having that job? (If you answered yes, please send your resume to 1600 Pennsylvania Avenue.)

The point is, the interpretation of facts often leads to unexpected results, and therefore even the most passionate pursuers of truth may find themselves frustrated. When you become Chairman of the Federal Reserve Bank, it is easy to understand why nothing is more important than your credibility and that you fervently uphold the trust placed in you and your office.

Let's jump to another extreme, and infamous, case—since it's the easiest way to see this abstraction in action.

After many months of hearing good things about Artie and his investment firm from a number of your friends and colleagues, you finally have the opportunity to have lunch with Artie. He's a nice guy, the wine is good, the filet is done the way you like it, and you find that each of you is respectful of the other's good manners, family values, and outside interests. During the course of the conversation, you ask Artie, "How's business?" and Artie replies, with a smile, "Never been better." That's reassuring, of course, because you're already predisposed to invest in Artie's enterprise, and a few days later, that's just what you do.

Now, the beauty (or the ugly) of this situation is that Artie has been absolutely direct, honest, and truthful with you. For Artie, business has never been better, and it's been getting steadily better as he's been having lunches with people like you. He's had numerous successful lunches with people who walk away and, a short time later, put their money into his enterprise. In fact, as you later learn to your dismay, that's the essence of Artie's business: He gets people to put money into his company that he then pays out to other people. The more successful lunches he has with others, like the one he just had with you, the better his business does. Business is, indeed, great, but it just so happens that the business is a Ponzi scheme.

This is one of the problems with getting at the truth. It's not always the answers you get that matter. Often, it's the questions you ask.

On the other hand, it would be helpful if Artie just told you what he means when he says "Business."

What Is a Business, After All?

Through the centuries of American history, business has been a dominant influence. In the process of dumping Royalty—as the Founding Fathers did so effectively and with such determination— we effectively discarded a venerable system of hierarchy-though-primogeniture that had served society very nicely for many centuries. If the king is no longer a kingmaker, however, then who (or what) is going to determine status and decide who holds the reins of power? Almost at once, from the time Europeans, escaping from Royalty-based governance, landed on this continent, the nation was more entrepreneurial than seigniorial in its leadership and social roles. The big farmer, the prospering merchant, the efficient manufacturer, and the large plantation holders became, in terms of reign, the kings, queens and lords. Business, not family, ruled.

The upside of this, as we well know, was the much-lauded ability of the members of the lower classes to rise to greater heights by dint of hard work, determination, and ingenuity. Business, in this country, has never been a second-tier enterprise. (For an English baron, dabbling in business might have been an acceptable hobby, but if it became a necessity, it was a sign of desperation.) The heroes, in the American iconography, were boys in Horatio Alger stories. Their rise from lowly bootblack to prosperous members of the middle-class perpetuated the sunny myth which was then embraced by an expanding population with emerging ambitions.

But in the newly invented society where power was a money-given rather than God-given right, there were none of the formalized, traditional restraints that sometimes helped the ruling class to remember to mind their manners and show respect for human dignity, even if they didn't have to. "*Noblesse oblige*," when observed and practiced, was a voluntary code of decency where no other code applied. *Noblesse oblige* was not comprised of laws enforced by bobbies with truncheons: rather, in families blessed with education and wealth it was an assumption about how people were supposed to behave toward each other and toward their underlings. This code of conduct suggested that one was supposed to behave with respect to everyone else, from tradesman and cook to peasant and beggar. *Noblesse oblige* did not make the haughty any less so, nor did it even the playing field with the hoi polloi. But it did force its adherents to make distinctions between behavior that was considered honorable and dishonorable, generous and cruel, tolerant and intolerant, gentle and brutal. Where and when it was respected, good things could happen, even across class boundaries. Given the luxury of inherited wealth, the families practicing *noblesse oblige* could pay attention to the quality of life rather than fret about matters of income.

The Luxury of Morality

But throw out the bathwater, and sometimes the baby goes with it. The soon-to-be iconic Horatio Alger hero had neither the opportunity nor obligation to pick up the finer points of seigniorial responsibility. His life was one mad scramble to get from the bottom to the top. By the time he had conquered his personal Everest, it was assumed that he pretty much had carte blanche to do what he wanted with his wealth. As a decent American boy of high standards and good principles, he was expected to stick to the principles of hard work and honest application which had made it possible for him to have this peak experience in the first place. But that did not mean the Horatio Alger hero would necessarily set up hospitals and libraries for his workers, or provide them with a post-retirement annuity, or even be polite to his driver. With a new definition of *noblesse* in this new country, *oblige* would have to be reinvented, pretty much from scratch.

And so it was, in a sense. But first there was hell to pay. Folks like the Carnegies and Rockefellers, the new royalty, littered their ambitious rise to wealth with the worn-out bodies of many thousands of families. Only at the peak of their wealth acquisition did *oblige* kick in. In pursuit of success, the new royalty broke up strikes, pounded heads, and destroyed lives. Hordes of immigrants were employed to dig, plow, and hammer for a pittance in their efforts to build America's infrastructure. Many immigrants eventually won access to basic rights of citizenship, education, and healthcare, but not until the nation was built upon their back. It took a bloodbath of a War and a storm of Civil Rights actions, but eventually slaves too won similar rights. Needless to say, the give and take between politics and business, populism and sectarianism, do-gooders and their opponents, was a long and often bitter struggle. Eventually, the new royalty helped to create libraries and medical buildings and gave bequests to institutes of learning that were entrusted to generations that followed. And the great enterprising capitalists of America—more dynamic, and often more reckless, than anywhere else in the world—accepted that certain rights of property, dignity, health, prosperity, and security were indeed required by those who worked for them. Starting, as it did, from zero, what has emerged as a kind of *oblige* of business has turned out, in the long run, to not be so bad.

Which is, in its way, remarkable, considering that what we call for-profit businesses really have just one operating principle: to make a profit. True, the goal, mission, purpose, or philosophy of the business may be stated in more salutary, and for a general audience, palatable terms. Frequently there are mission statements (particularly popular now) that emphasize the prime importance of serving customers, bettering society, delivering a certain style or quality of goods, or in some way providing other great benefit to people who pay for services. And, indeed, such high-minded statements and beliefs can be inspirational as well as motivational. But when a for-profit business can no longer make a profit while pursuing its stated goals or missions, be assured that either the business will cease to exist or the mission will change. There's nothing in the nature of for-profit business itself which implies or states that it need perform any public or social good.

Noblesse oblige is a wonderful term for those who are financially secure, but let's face it, a lot more difficult if you're in straitened circumstances. In other words, those desiring to do good, must first do well. I need to be able to take care of myself before I can take care of others. More broadly, this raises a curious question about morality itself: can I afford it?

While we probably don't like to admit it, Morality is not universally affordable.

The Question of Affordability

Is it wrong to steal? You say it is? Okay, well let me pose this question: Would you steal a stale loaf of bread to feed your starving young children? Would the theft be less wrong because the need is so profound? Ethicists don't like the notion of "more wrong" or "less wrong," and we'll defer to their expertise. But separate from the ethical debate is the affordability issue.

If you walk past the panhandler with coins in your pocket, you may or may not decide to give him the money. You may suspect that he'll just buy booze, and that your donations will be more effective if made to the Salvation Army. If you have money in your pocket, you have a choice. If you don't have money in your pocket, there's no choice, no matter how compassionate you may feel.

Here's another one: In the middle of selling your company, you realize that you're receiving a full price for a couple accounts that are being reviewed and may leave in a few months. If you need every last penny from the sale for your very survival, you may not mention the fragile accounts. On the other hand, if the additional $20,000 in proceeds is nothing more than a rounding error to your already massive fortune, you can easily and affordably take the moral highroad and be forthright.

On a larger scale, a developed society may choose to ensure that the poorest, most vulnerable citizens will not go without shelter, medicine or food. While it may be a noble, laudable condition, it must be afforded. If you undermine the economy, the social benefits it provides cannot continue. If social benefits continue to be provided beyond their affordability, they must be paid for through borrowing. As borrowing increases, debts grow to exceed the borrower's ability to repay it. This results in defaults that leave the lender unpaid (and unlikely to lend again in the future) or debt that must be transferred to obligate another person or entity.

Yet, a certain *oblige* has, indeed, and in spite of all resistance, grown up around the ethos of American business. Today, corporations of all sizes approach business with an extraordinary number of "shoulds" that do not seem, on the face of it, to relate directly to the *big* should (...make a profit). Many business owners today believe they should compensate workers above minimum wage as well as pay them for overtime. Whether or not they are legally obliged to do so, many employers feel they:

Should provide decent working conditions;
Should give time off;
Should help make arrangements for retirement benefits;
Should be aware of the environmental impact of their companies;
Should not discriminate;
Should report data in a fair and accurate manner to shareholders;
Should not engage in bribery.

We deem that some of these "shoulds" are just. Others, necessary. Others, onerous. But the point is, for-profit business in America is not, nor has it been for the past hundred years or so, a freewheeling game where you can rake in the bucks without regard for the consequences. A certain *oblige* has settled into this terrain, and no matter how pesky or onerous the "shoulds" have become, they cannot be eliminated without setting back American business a hundred years or so.

To a large degree the *oblige* is enforced by norms of organizational behavior. Get a job at any reasonably well-run business in America today, and you will find that you are paid the agreed-upon wages, that you can still cast secret ballots for candidates of your choice, that you work in a sufficiently safe environment, and you are treated with a reasonable amount of respect. If you are a customer dealing the same companies, you have confidence in the contracts and agreements you make, and you also have reasonable expectations about the delivery of goods or services. If you are a principal in such a business, you proceed under the assumption that your business can't be arbitrarily shuttered without cause, that your right to make a profit will be respected, and that you will maintain control over disposition of company assets. Note that all these assumptions are, in fact, the result of a wide variety of social and legal pressures and can only be maintained so long as the checks and balances are in place to preserve respect for the fundamentals. They are a kind of "base level" of trust built around concepts such as the right of ownership, regard for human rights, and trust in an effective legal and judicial system.

None of these fundamentals can be taken for granted. They are the values that make prosperity possible. These fundamentals also create conditions that motivate people and allow them to believe in the possibility of a better future.

Bully or Builder?

Of course, this higher-evolution view of American business omits a lot of rotten eggs. Al Capone ran a prosperous American business, and so do a number of Mafia families. But, obviously, we like to make some distinction between the accepted standards of legitimate business enterprises and the strong-arm tactics of those who pursue their ends in shadier alleys. The freedom of the American press has played a role in this awareness-making process, distinguishing the legitimate from the non-legitimate, the builders from the bullies. So we tend to have a fairly high level of confidence when we sit down with a guy like Artie—who has been in business many years, comes highly recommended, and appears to have sterling credentials. In the absence of evidence to the contrary, we assume Artie respects the way things are supposed to be done. And that, in a nutshell, is why we don't ask him many tough questions. We feel we don't have to be that thorough. He has a thriving business in a country where laws are supposedly enforced; where property rights are supposedly respected; where representations made on behalf of a corporate enterprise are supposedly legal and accurate. As we dine in style with Artie, we assume, a priori, that this guy is in business to make money, but we also assume, a priori, that he does not steal personal fortunes, that he abides by contracts and employs conscientious professionals. We assume he is considerate of the long-term prosperity of his customers—in a word, that his profit-making activities adhere to some accepted social norms. You assume that you can trust him. And you want to trust him. You want the returns to be real, and you don't want to ask the rude questions that might break the magic spell.

But what if you *can't* trust him?

What if Artie, like old Bernie Madoff, has been getting away with a vast Ponzi scheme? What if, at the end of your comfortable lunch with him, you put the yolk of your nest egg into his fund—only to see it scrambled, along with some $50 billion of other people's money? If that happened to you, how much would you continue to trust the business professionals you deal with in the future? How would you change your underlying assumptions about the operating principles?

Perhaps you'd ask a lot more fundamental questions, but would you be able to get answers you really need?

This, it seems, is the specter of doubt and distrust that overhangs American business in the wake of the market crash of 2008. Perhaps there were only a few Ponzi-schemes like Artie's (Bernie, Robert Miracle, Scott Rothstein, Allen Jacobson, to name a few). But at the same time a lot of us were sitting down to lunch with a lot of individuals representing American enterprises in which we had faith. And because we made assumptions about their reliability and trustworthiness and so forth, and because (having made these assumptions) we didn't ask the right questions, we found ourselves in a heap of trouble. It turned out, as we later learned, that a lot of banks and mortgage companies were lending to people who they knew would be unable to pay. Staid, respected financial companies with long track records of decent behavior knowingly purchased sub-prime mortgage-backed securities that were precarious to the point of worthlessness. The nation's largest and most prestigious insurance firms were insuring the value of these securities knowing full well they would be unable to make good on their obligations if the bubble in the housing market exploded. Banks with international reputations and global reach were carrying out transactions and were earning fees for services that, they knew, were little more than bets on proprietary checker games.

Yes, questions were asked. In fact, some people asked exactly the right questions. But their voices were drowned out by the spectacle of so many exalted institutions engaging in what appeared to be acceptable business practices. The sheer size of the swindle made it convincing. How could so many players, of such stature and reputation, be taking us all to lunch with the sole purpose of selling us on some extravagant Ponzi scheme?

Well, we know what happened. But we had, and still have, no idea what lessons we're supposed to learn. The big institutions were propped up. Play resumed, with some slight variations, and the shuttered homes and lost jobs became a political football.

But hope, in the American psyche, took a nosedive.

A Rising Tide that Floated Few Ships

One problem, of course, is that the majority of those involved in this ridiculously large-scale bonanza reaped enormous rewards they were permitted to keep. True, Bernie Madoff and a few others went to jail. But in aggregate, the majority of big players who enthusiastically engaged in, and were responsible for, the financial crisis, took home billions of dollars that became the working capital of their lavish lifestyles. The amassing of wealth in the upper strata of the American population did not begin with the mortgage-backed security debacle. However, the outrageous shows of largesse, even in the aftermath, puts the spotlight on those who took home multi-million dollar bonuses. And they were doing it while, elsewhere in America, countless properties were being foreclosed on, shuttered, abandoned, and even plowed under.

During the 2008 crisis, something seemed to have gone suddenly and terribly wrong. But the shift of wealth was not sudden at all. A steady gusher has been flowing toward the top few, in ever-increasing quantities, over a period of twenty years or more.

Perhaps systemic risk is harder to notice when the "greed-is-good" philosophy supports capital flow. There is something so piggish about the G-I-G ethos that it seems almost farcical. As the audience watching this farce, we assumed (how wrongly, it turned out) that the greedy grasping buffoon would eventually be clouted on the head by a bat-wielding administrator of justice. And as the buffoon fell to the floor, scads of money would fly from his pockets and fall into the hands of the hardworking, righteous villagers. But then, much to our astonishment, the greedy buffoons were never smacked down. *Au contraire*, after a few raps on the knuckles, they were picked up, put on their feet, and brushed off, whilst a collection was taken up among the populace to replenish their squandered cash.

This was a cause for outrage which prompted another kind of street theater: the Occupy Wall Street Movement.

Occupy *What?*

To everyone with a pen and/or a camera who interviewed participants of Occupy Wall Street (OWS), it soon became apparent there was no stereotypical participant. Sure, there were despairing, unemployed twenty-somethings with ripped jeans and bandanas waving signs and loudly protesting injustice. But their number included recent graduates of colleges and universities bearing debt-loads into the hundreds of thousands of dollars, ready to take almost any job anywhere that would give them some hope of a future income. There also were veterans of the social movements of the nineteen-fifties and sixties (now in their sixties and seventies), followers of the media and, in many cases, participants in earlier protests. All of them wanted to believe that this fresh movement could somehow awaken the titans of Wall Street to the crisis of trust or, at the very least, make them blush. All around the tents and tables and "Organize" territory, there were, of course, cops. But there were also employees of corporations large and small who had taken lunch-hour breaks or day-long journeys to witness...well, *something*. What was it they were participating in, sympathetic about, wary of? What were the Occupy Wall Streeters *doing?*

OWS captured enough headlines that it became a frequent topic for my on-air appearances. I didn't know much about these folks, so, as I have mentioned, I donned my jeans and sweatshirt, took my NBC Universal credentials and digital recorder and sat on the sidewalk in McPherson Square in Washington, D.C. to talk with them. Their answers were as diverse as the participants. We want the big banks broken up, some said. We want to end the collusion between big money and Washington, said others. We want the investment bankers brought to justice, said all. We want the fat cats to return the money they made off the backs of ordinary people, and we want them sent to jail. We want to change the system that puts wealth in the hands of the one percent and deprives the other 99 percent. We want our houses back and an end to foreclosures. We want money to go to education and health care, not bailouts and wars. We want the government to stop putting people in jail for victimless crimes. We want Wall Street to listen. We want Washington to listen. We want jobs. We want opportunities. We want to see some reason to hope that things will get better and not worse.

The OWS movement had remarkable magnetism; it drew attention and support from nearly every spectrum of society. However, critics and pundits were quick to assert that it seemed to be a movement that had no place to go. It was all issues, no solutions, they said. All complaints, no plans. All heat of passion, no rationality.

I asked every one of the protestors to tell me their three main goals. I asked what would signal to them they could end their camp-in and go home. Not a single OWS participant could answer either of those questions. Their goals were "change." And how would they know when that change had occurred and when they could end their vigil? They responded, "We'll know."

But to dismiss them completely was to misunderstand the clamor and misread the problem. There was, in fact, no shortage of proposed objectives and solutions. One frequently mentioned solution: Break up the banks that were "too big to fail" so we would never again be in a situation where the economy could be crushed by market exuberance. So the American public would never again have to bail out institutions teetering on the brink of collapse. Isn't that exactly what leaders in business and government called for during the financial crisis of 2008?

Another suggestion: Ease the pressure on homeowners whose mortgages were underwater through no fault of their own. Change the mortgage terms in ways that would allow them to rebuild their lives, recover from their losses, rebuild their neighborhoods, and move ahead. Isn't that what the rescue programs and "bank incentives" were supposed to do?

And another suggestion: Relieve people of the onerous burdens of student debt. As college tuition, student loan amounts and loan interest rates were climbing, fewer and fewer graduates were able to find any job that would allow them to pay off their student loans. Every lawmaker, executive, and citizen seemed to be in favor of affordable education, but the definition of "affordable" ballooned many times beyond inflation

What most characterized the clamor of voices in the Occupy Wall Street Movement was not the radicalism of their views but the fundamental ordinariness of what they were asking for. They asked for limits on executive compensation for tax-payer supported businesses—was that so outrageous? They asked for restraints on the influence of big money in Washington. Not exactly a barn-burner. What about *not* giving huge bonuses to Wall Street titans who had driven their companies into the mire and were subsequently rescued by bailouts? Was that really so much to ask?

The requests, voiced as demands, were not at all excessive. In fact, they echoed the uneasy questions that were being asked and the gripes that were being shared in homes, businesses, and town halls throughout the United States. It really was not difficult to identify what was wrong, what decisions needed to be made, which objectives to reach. The problem was, *nothing was done.*

The economy had come close to collapse. The hurt was still being felt—more profoundly than ever—throughout America. The malaise of the Recession was likened to that of the Depression, and the calls for change could be heard across the land. And yet the political and economic scene seemed to be in a state of frozen animation. Over all, and worst of all, was the growing dread of imminent peril. The more things did not change, it seemed, the greater the likelihood, even inevitability, that it could all happen again. The OWS protestors were losing hope and felt that the American Dream wasn't possible for them. They didn't think they stood a chance because they could no longer trust that a level playing field existed. Without this hope and trust, we risk losing more than just our economy; we risk losing our American way of life.

In 2008 we were sternly reminded by our business and political leaders, who spoke with remarkable unity, that the economic events leading up to that year had brought the nation to the brink of financial collapse. Gathered together in Washington, grim-faced and stalwart, leaders assured the American people that the only lifeboat available to us was floated by the U.S. Treasury. They prophesized that without the bailouts we would all go down. We listened, stunned. We heeded the words of warning and accepted the solution (we had no choice). We hearkened as they proclaimed "never again."

And the result...

Even larger bonuses. The frightful expansion of the biggest banks and the increase in their power. The failed rescue of homeowners. A stock-market surge that seemed unhinged from fundamentals. The increased concentration of wealth in the top one percent of the one percent. The continued diminishing of educational performance and increased difficulty of most people trying to afford it. The growing numbers of incarcerated Americans. The ever-widening opportunity gap between those born with great wealth and the rest of society. It was as if, having narrowly missed the first iceberg, we turned the ship around and tried to hit it again.

Tea Parties

While the Occupy Wall Street movement sought a coherent message to unite its front and its followers, the Tea Party never seemed to have that problem. Give us lower taxes, a balanced budget, and less spending! It was their mantra, their motto, their policy, and their battle cry.

To a Tea Party united by a single, united idea, the solution appears obvious: deprive the multi-headed hydra (big government) of its sustenance (the people's money), and its multiple tentacles (overreaching programs and policies) will surely wither and die. Like a decaying artifact, government will gradually cease to have the power to meddle in the lives of ordinary Americans. Tea Partiers believe that through these means we can restore the treasure-trove of freedoms that our founders intended for us.

It's nice to have a solution, especially when it serves as such a singular rallying point. The lower-tax credo had a far more immediate impact on Washington legislators than the various outpourings of complaints from Occupy Wall Street. In media analyses, of course,

OWS is clearly left and the TP is clearly right. And, indeed, the Occupy Wall Street message has been more welcome among Democratic voters while the Tea Party platform has been largely generated and supported by the farthest right wing of the Republican Party. But if we strip away Right and Left labels for a moment, is there something that both movements have in common?

One thing we know from the chronology of events is that the financial crisis and the way it was handled by leaders of business and government provided a catalyst for both sides. For those joining in Occupy Wall Street, the most egregious sins were those committed by Wall Street firms, big banks, and the moneyed interests in Washington, D.C. For Tea Partiers, the sin was overreaching big government, and the bailout of Wall Street firms.

But the OWSers and the TPers aren't one-hundred-percent opposed. They agree that something isn't right and that something needs to be done. They agree that the system of government dreamed up and codified by extremely smart and imaginative white landowners, statesmen, soldiers, farmers, and politicians has not evolved to meet the challenges of a vastly changed and technological world. They agree on the need for accountability and consequences. They both demand a shared vision and goal of American excellence and a commitment to high and noble ideals.

One rather despairing view is that if we could only tap into the brains of our forefathers and understand what they intended, and act in accordance with those desires and constraints, today's problems would be solved. Another no less despairing view is that the operating principles which have been serviceable for nearly 250 years can be dumped and replaced by a brand-new framework of operating procedures. But somewhere in the middle of these widely divergent views is a shared longing for things to get better.

But what is "better," and how do we get there?

The following are questions over which Left and Right get into shouting matches:

    L: "Better" means regulations imposed on big business so we don't get screwed.

R: "Better" is a free market without government meddling.
L: "Better" is a safety net for the indigent and incapacitated.
R: "Better" is letting people help themselves and help each other without mandates from Washington.
L: "Better" is free education, free health care.
R: "Better" is letting people keep what they earn and spend it as they wish.
L: "Better" is laws that punish social discrimination, political corruption, and economic exploitation.
R: "Better" is a nation with the fewest laws to interfere with individual rights.

But Left and Right are not making much progress by yelling at each other. So where is the common ground? Is there any?

If we view the financial crisis as the spark igniting both forms of protest, we may indeed see some common ground in the common sources of shared frustration.

When Sparks Fly

Let's go back to that ghastly moment when the economy was on the brink of collapse. The blow might have been less brutal if our rescuers weren't the very same people who were burning the house down. When a house is afire and the firemen arrive, you hope that trained experts are about to bring up the big hoses and save the occupants. But in the case of the financial crisis, the people arriving at the scene to judge the magnitude of the fire were, themselves, the arsonists. They had fueled the blaze by making us giddy with their promises of new furnishings, new additions, and endless expansion. And now, quite suddenly, they were telling us that forgotten embers in the floor beams were alight, threatening the entire edifice.

Oh, they tried to change uniforms quickly, getting out of their arsonist togs and into firemen's uniforms. And they put on quite a show at the main firehouse in D.C. But what they told us, wearing the stern visages of seriously worried elders, was the tale of a fire long simmering, now burning out of control. And the very people who started the fire said they would take care of it.

In the meantime, what did our leaders expect the rest of us to do? Nothing. Absolutely nothing. We were asked to stand behind the police tape and watch.

The feeling of helplessness got worse over the next couple of years. All Americans felt the impact of rising unemployment directly or indirectly. If you weren't unemployed, then perhaps your brother? Your friends? Your colleagues? You watched the equity in your home diminish, or watched the investments set aside for a parent's care (or your own future!) turn to vapor. Consider the helpless feelings of a student with $80,000 in loans who is unable to find an entry-level position. Consider the feelings of a homeowner in a neighborhood where homes are being shuttered and vandalized, or those of a construction worker whose hours (and pay) are cut by half or two-thirds, or whose job is completely gone. Left, Right, OWSer, TPers, all have shared, directly or indirectly, in these experiences.

We were lied to about the nature of the prosperity, we were lied to about the nature of the rescue, and now we would be lied to, again, about the repairs being made. But we also lied to ourselves. Like purchasing the lottery ticket, we suspended disbelief and ignored contrary evidence. And we believed. We believed because we wanted to, and it was easier than dealing with reality. We trusted, so we borrowed, but we also borrowed because we made ourselves believe in the magical conditions we greedily desired. We were human. We made mistakes. But that means we are absolutely jointly liable for the consequences.

In 1980, U.S. average household debt was 55 percent of annual household income. In 2009 household debt ballooned to 110 percent of household income. That's the data for average households, and "average" includes a hell of a lot of us. We used credit cards. We took out mortgages, home equity loans, boat loans, and car loans. We did it all. If you want to blame Wall Street and the banks, go ahead. But don't stop there. Don't stop until you get to a mirror and shake a disapproving finger at yourself. We are all accountable. So part of the solution to this mess is looking beyond someone else to blame and conducting an honest self-check.

But it was the big liars who became, for a time, the assassins of hope. They didn't set out to burn down the building. When the embers burst into flame, they were actually scared (even those who had seen it coming). They didn't want the furnishings (especially *their* furnishings) to go up in smoke. And some truly felt they performed heroic deeds during the so-called rescue. What they didn't anticipate, however, was the lack of confidence that would result from their behavior. They didn't expect there would be tents planted in public parks with people yelling, "We don't trust you," and tea-party gatherings in living rooms around America with other people yelling the same thing, "We don't trust you!"

A lot of expectations have been shattered. Hope has been betrayed. And if they are to restore the trust and hope, the assassins will have to, somehow, change their ways.

Contributors to Chaos

What is it, exactly, the assassins *do* to diminish aspirations, crush dreams, and stomp out the glowing embers of industriousness, creativity, enthusiasm, and hope?

Imagine a teacher and students in a classroom that is custom designed to instill hopelessness, dissension, and chaos. Several innovative methods are used. First, the rules are said to apply to everyone, but there are always exceptions. How are the exceptions chosen? As randomly as possible, but with enough frequency to keep everyone on their toes. Today, Johnny will be allowed to leave his seat at anytime to get a drink of water. Tomorrow, he'll be punished for doing exactly the same thing. For a couple of days, Janet is permitted to talk to her neighbor, but on the third day she is given a detention for the same behavior. High standards are set for classroom work, but as the students soon come to realize, not everyone has to meet those standards to get a good grade. Every once in a while, the teacher gives an F to someone who writes a perfect paper. The teacher occasionally recognizes when a student struggles with a difficult test, and he gives the student an "A,"despite having earned an "F". But only sometimes. And if the students question the teacher's actions, he points out that he is the teacher; and therefore, he can do what he wants.

With the confusion caused by ever-changing rules and ever-shifting standards, students learn to curry favor with acts that tip the odds their way. Charlie starts leaving M&Ms on the teacher's desk. Charlie soon finds that his grades improve without any extra effort. But his efforts are soon outstripped by June, who discovers her teacher's weakness for Mars Bars. Charlie's grades fall, June's rise. Before long, to their mutual dismay, they are overtaken by Simon, who seems to have an endless supply of chocolate truffles. The news is soon out, Simon can do no wrong. His position in the class is unassailable. (In Washington, this is called lobbying.)

Consider the hope and expectations of the students in the classroom I describe. They learn their chances of success depend on one thing—how many and which kind of chocolates they bring to the teacher. And they learn if you can't buy chocolates, you don't have the remotest chance of being a winner. Without chocolates, all the hours spent studying and all the earnest class participation mean nothing. If you can't figure out a way to get high-class chocolates into the teacher's stash, you might as well not show up. You don't stand a chance.

Through repeated trial-and-error and with frequent discussion among themselves, the students start to see the realities of their situation. The general mood of optimism is replaced by bafflement, then skepticism, and finally pervasive cynicism. The natural hard workers slog on because that's what they do. But they do so with growing feelings of despair as they see other students—including the bearers of M&Ms, Mars bars, and truffles—getting far better grades for doing far less work. Students who began the year with a can-do attitude find they can't quite suss out the system or come up with the right set of goodies. They lapse into a state of inertia viewing the motivated hard workers as suckers and the truffle-bearing students as kiss-asses. As for the students who are seriously involved in chocolate-bearing, their lives become ever more frenetic as new rivals emerge, bearing ever-more-expensive gifts to sweeten the tooth of their teacher.

Let us note, however, that the chaotic nature of the classroom, the multiplicity of contradictory drives and motivations, and the growing distrust and resentment among students are matters of little consequence to the teacher. He is happy and well fed. In fact, he grows more cheerful by the day, even as incipient diabetes sets in.

This is a hope assassin, and this is how he operates. His once-eager students are reduced to a scrappy, factionalized, and resentful bunch. Each operates independently, each with his or her own cynical formula about what must be done to survive and thrive in this type of environment. Previously, students would become better citizens by earning better grades. Under the new system, once-healthy competition is reduced to a joke. By failing to reprimand the chocolate bearers for their bribes and failing to put a categorical end to the practice, the teacher sends a clear message. Favoritism is not only tolerated, but encouraged. A general pall falls over classroom proceedings. Classes continue, but in a kind of daze. Any chance of achievement seems dubious. There is simmering resentment at the injustice being served. And when, at the end of the year, the chocolate-donors are the valedictorian and salutatorian, the class surliness finally manifests in the form of sign-bearing and sit-ins.

But…there are those who ask, what are these students complaining about? Why don't they just get themselves some chocolates and play the system?

Chapter 3

The Psychology of Reward and Punishment
"The Pursuit of Happiness"

Sometimes, rats respond in ways that we don't expect.

It was a Friday afternoon and Burrhus Frederic Skinner, an intrepid explorer of psychology, was in his lab preparing for the weekend's experiments involving his favorite subjects: rats. Skinner set up the kind of reward system that would operate automatically in his absence, recording the animal's response. If the rat pressed a lever, a food pellet would emerge. The rat could "voluntarily" decide to push the bar or not. But if it kept being rewarded, in time it became "conditioned" to repeatedly press the lever to get its pellet.

Skinner had also explored another aspect of this—extinction. What would happen if there was repetition without response? How long would it take to weaken the response that had been strengthened by reinforcement (repetition with the expected conditions)?

On the weekend in question, Professor Skinner made one of those accidental discoveries that are so fortuitous in science. He was short on food pellets and realized that if he was going to have enough for the coming week, he would have to spend the entire weekend making them. Instead, he set up his experiment so that the rat would only receive a pellet every three or four responses. In other words, there would eventually be a "reward" for the rat, but it was a lot less predictable than one-push, one-pellet.

What Skinner discovered that weekend was the power of "intermittent reinforcement." The animal made *more* responses in extinction than it would have if it had received a reinforcement (food pellet) for every response.

Skinner felt he had made a crucial discovery. If animals responded this way to intermittent reinforcement, what could this tell us about human responses? If repetitive behavior always produces the same result, doesn't that get a little boring? (Ho-hum—I'll press the lever again and again I'll get a reward.) On the other hand, if I press the same lever over and over again, and nothing ever happens, I'm sure to give up after many tries (the extinction response). But what about this situation where I keep trying and trying, and…every once in a while…Jackpot! The coins roll out of the slot machine and I'm awash in feelings of euphoria. Intermittent reinforcement is a powerful inducement to action.

Discouraged Rats

How do we respond when we're confronted with adversity? When a proffered reward, consistently delivered, is withheld (either arbitrarily or intentionally), how do we react? Do we blame others, ourselves, or the gods? Do we lapse into a kind of torpor, incapable of response until prompted by new stimuli? And most interesting of all, are there some humans, dogs, rats, or pigeons that easily give up hope while others persist? Do they believe with blind optimism that if we keep trying and trying, we'll eventually get results?

Psychologists during and after Skinner were quick to ask these very questions. Many of the assumptions about the bases of operant conditioning have since been re-studied, modified, revised, and refined.

Take, for instance, the ineffable qualities of personality and character that we feel instinctively in the presence of other people. Just as there are personalities gregarious and shy, flamboyant and modest, outgoing and reserved, so too are there people whom we instinctively label as optimists and pessimists. You know the optimists because when the food-lever doesn't deliver, they keep trying, then go looking for other food levers, and finally climb out of the box and go hunting for some reward other than the unreliable container of pellets. You know the pessimists because they find the problem with the malfunctioning lever very discouraging. With each try, their indifference grows until they retire to a corner and wait for someone or something to come along to improve the situation and end the famine.

Obviously, these differences in attitude make the whole process of operant conditioning far more complex. How do we understand what distinguishes the optimist from the pessimist, the person who retains hope—despite adverse conditions—and the one who loses hope even when viable alternatives are within reach? Is the optimist inherently more imaginative and creative, and therefore better able to find his or her way out of a ticklish situation? Or, conversely, is this the less-imaginative person, the one who has stubborn will and barges ahead, overcoming all obstacles, simply because he or she does not have the intellectual scope to see the formidable obstacles that lie beyond? Further—are the differences environmental or genetic, a product of nature or nurture? And perhaps most important of all, at least from a social perspective, can a dyed-in-the-wool pessimist undergo a change in outlook or circumstances that will remold his or her psyche into that of a well-functioning optimist?

Among the post-Skinnerian psychologists who picked up on these questions were Martin Seligman and his fellow researchers at the University of Pennsylvania. In his clinical work, Seligman questioned the process of therapy that had prevailed since Freud's day. Certainly, he said, there was some therapeutic value in exploring a patient's past and discovering the sources of anxiety and (to use the Freudians' favorite term) "neuroses." But what were the outcomes of such therapy? How did it help ordinary people cope with problems, get a grip on themselves, and face the challenges of daily life?

What emerged from Seligman's research and clinical work became the basis for a new approach loosely categorized as "positive psychology." Though the very phrase evokes images of rah-rah pep talks and rousing sermons from inspirational speakers, Seligman and his students were far more interested in the scientific underpinnings in the reinforcement of positive actions, attitudes, and approaches. As he describes it, his goal was not to toss out traditional psychology but rather to "supplement its venerable goal with a new goal: exploring what makes life worth living and building the enabling conditions of a life worth living." Their findings led them to the conclusion that consistent reinforcement of a positive outlook produced far more than a temporary change. As detailed brain research began to catch up with clinical observations, Seligman was able to show that consistent reinforcement of positive actions, thoughts, and behaviors would eventually produce a discernible transformation in the synaptic structure—the flow of electrical pulses through neurons in the brain.

Stepping back, what Seligman explored was the pleasure principle—that is, our innate desire to avoid pain and seek pleasure. But in Seligman's research he sought to explore the many complex manifestations of human gratification that can be classified as pleasure. In his book *Flourish,* Seligman summarizes it this way: "Positive psychology, as I intend it, is about what we choose for its own sake." Of course, what we choose for its own sake could be as attainable as a back-rub or as arduous as memorizing all of Shakespeare's plays. In *Flourish,* Seligman proposes there are five different elements we choose for "their own sakes"—positive emotion, engagement, meaning, positive relationships, and accomplishment. Seligman gives a detailed and scholarly analysis of each of these elements, but what it comes down to is defining and understanding *what we want.* These elements are:

Positive emotion—feeling good about the back-rub.

Engagement—Getting so caught up in an emotion or experience that we go into a kind of timeless "flow state."

Meaning—Belonging to and serving something that you believe is bigger than the self.

Positive relationships—Being with people you like, love, admire, or just want to hang out with.

Accomplishment – Winning, achieving, mastering a skill, climbing a mountain.

As Seligman suggests, it's likely we all need some combination of these elements to feel a sense of happiness, fulfillment, or well-being. But of course, for each of us, the priorities are weighted differently. For some, the strength of positive relationships will readily compensate for a dearth of accomplishments. For others, winning matches or mastering skills are the pinnacle—other factors matter less. But all of us recognize that our lives will be improved if we can follow avenues that lead to some of these qualities of well-being.

But seeing what we want, and getting it, are two different things. The followers and practitioners of positive psychology have marshaled convincing evidence. They have shown that with appropriate intervention and periodic reinforcement, we can enjoy the results of therapy that has indelible effects on the paths of neuron activity, which lead to more affirmative thoughts and actions and an overall more rewarding view of life. But, obviously, we don't all have access to resources that we would need to make the most of positive psychology. For the rest of us—that is, most of us—some other element is needed. That element is hope.

What *Is* Hope, Anyway?

This is a question that psychologist C. R. Snyder, a professor at the University of Kansas, attempted to answer. In *The Psychology of Hope* he recounts the experience of walking into his academic library to begin research on the topic, only to find that nothing was available. While hope looms large in literature, history, religion, and the survival and renewal of civilizations, it has been omitted from the working vocabulary of clinical psychology. Professor Snyder set out to remedy the situation. He began by looking outside the library, by observing people and asking them about hope.

He discovered that hope is frequently given a bad rap: "The view on hope appears to turn on whether it is perceived as being realistic. Questioning whether hope is built on anything substantial, many have viewed it as a curse. That is to say, hope is portrayed as an illusion, totally lacking a basis in reality." In this view, hope is something for dreams and fantasies. It's the nebulous wish for something far off and unattainable. But as Snyder launched into his interviews and began to identify what people really mean by hope, he found that it signified something quite different. "I join recent social scientists who suggest that hope involves the perception that one's goals can be met. It is how we think about reaching those goals that provides the key to understanding hope."

In other words, far from being a vague, illusory longing for something-or-other, hope is believing we know how to get from point A (whatever it is) to point B. To make this journey, Snyder concluded, requires a few "basic mental components—goals, willpower, and waypower."

Defining goals is one of the "basics" of strategic action. But when we are full of hope, goals lie somewhere between the highly pragmatic and the unattainable. For those who are literate, wanting to read the morning newspaper isn't a far-off hope. But for someone who is illiterate, the reading of any newspaper is an attainable but high goal—well-qualified as a *target* for one's hope. If you are fully mobile, it makes no sense to say, "I hope I can walk across the room." Yet for someone learning to walk with a prosthesis, the hope of getting up and crossing the room may be so challenging that it can hardly be imagined. The goals involving hope, Snyder concluded, "fall somewhere between an impossibility and a sure thing."

What about the role of willpower? By definition, it means the will to do *something*—and certainly that vital force is stronger in some people than in others. But as Snyder sees it, willpower is also a directional force. It's the vibrancy and strength that helps propel us from point A to point B—and, therefore, willpower is linked to getting somewhere definite. Although there are a wide range of impulses and stimuli that can get us moving, willpower "reflects our thoughts about initiating and sustaining movement toward desired goals."

Less familiar is what Snyder identifies as "waypower," what he calls "the mental plans or road maps that guide hopeful thought." Given that you have reason to hope you can play the piano (a goal) and you are determined to do so (willpower), it's likely you'll need to get some music, find a teacher, and figure out how you can get the daily practice you need to become reasonably proficient. And that's where waypower comes in. As often as not, waypower involves the ability to make initial plans and also make course adjustments in order find one's way over and around obstacles. But whatever the final route, we need the ability to shape and reshape plans in order to reach goals—and that's waypower.

"Simply put," Snyder concludes, "hope reflects a mental set in which we have the perceived willpower and the waypower to get to our destination."

Hope vs. Wish

As an example of hope in high gear, the Mega Millions Jackpot that I describe in Chapter 1 may be somewhat lame. Going to your nearest convenience store to buy a lottery ticket doesn't seem to require a lot of willpower or waypower, even if the destination (that Jackpot) fits the definition of being just barely attainable. But on the other hand, maybe it's not so far off. Getting a few bucks together and parking yourself in line for a ticket does, in fact, require a certain kind of willpower. As long as you're buying a legitimate lottery ticket at an authorized retailer, you've fulfilled the criteria for waypower (yes, with the right number, you could become a winner). And whether the destination is realized or not—whether you are in fact the recipient of the multi-million-dollar jackpot—no one would ever doubt that you have a destination.

But what happens when you lose—as millions of people do every time? This is when we begin to think about the differences in human nature that Seligman, in his research, has found to be so revealing. There are some of us, let's face it, who will go into a blue funk when the Lottery winner turns out to be *not me*. Others, in the hope-springs-eternal crowd, are already in line for the next Mega Millions Jackpot. And then there are some—shall we say the exemplars of positive psychology—who will begin to think, "Hmmmm, maybe standing in line every week and buying a lottery ticket is not the only way to reach the destination I'm after."

It's this last group I suspect most of us would like to join. This is the Gang of Hope that, having established a goal, has not only the willpower but the flexibility to try a wide range of means to attain that goal. Let's take, for example, a studious and inventive kid who works hard in school, applies himself to his studies, and shows a native aptitude for problem solving. His goal is to get into a very good college and to graduate—obviously because, thereafter, he expects to have great opportunities. He succeeds in getting into Harvard. Hooray. Then something very weird happens. What's going on at Harvard doesn't interest him that much, and the goal of simply graduating begins to pale in comparison to some other interests that he has. His destination is no longer a degree but…something else. He drops out of Harvard. But there's no staying the willpower of this guy, and before long he's finding ways over and under and around all obstacles to bring the world an innovative operating system that can be applied to the nascent generation of personal computers.

Okay, it may seem like cheating to use Bill Gates again as a shining example, and I'm perfectly aware there are many who may resent his billions. But my point is, he couldn't have done it without hope. Yes, it's the same ephemeral quality that compels some of us to buy tickets for the jackpot; compels others to take a course that will lead to certification for jobs they want; others to make investments they believe will pay off; others to work long hours in labs, to invent patentable products, to start a corner store or online business, to get their kids dressed and put on the school bus every day. With hope—willpower, waypower, and a destination—we devise ways to get homes and cars and goods, to help other people, to form organizations and propound causes, to create works of art, to chip stone for cathedrals and weld beams for skyscrapers. All in all, it's a damn good strategy for getting things done. A way to get there from here. And a crucial ingredient in positive outcomes.

Fortunately, we are part of a culture that has a tradition of putting a premium on the hope of individuals. That thing called "American spirit" was emulated by freedom-seeking religious iconoclasts, pelt-seeking beaver traders, land-seeking farmers, and gold-seeking miners. It crops up again and again in the writings of Bradford, Franklin, Hamilton, Jefferson, Thoreau, Stowe, Melville, and Twain. The individual right to pursue Happiness is written-in, as a goal, in the foundational documents of our nation, and it's referred to, repeatedly, not only in literature but in legal judgments of the past couple hundred years.

What it comes down to is this: we can deal with intermittent rewards. If we push the bar and no food pellet emerges, we are willing to keep trying. Unlike Skinnerian rats, however, we can also find alternative ways to get what we want. Inherent in our laws, beliefs, and traditions, we have the right to exercise our willpower and waypower, to jump out of the box and find other sources of food pellets (even organic ones). To repair the broken-down mechanism ourselves (even if it means calling in the IT people). To overcome frustration, starvation, and victimization by repairing and, if necessary, replacing the power guy in charge of this experiment (even if it means sweeping out all the incumbents). We have rights not only to have positive thoughts and intentions but also to implement the strategies needed to get ourselves out of the box. We have the right to hope.

But It's Not Inalienable

As we're keenly aware, however, all it takes to destroy hope is a human monster.

It is remarkable, when we think back, how America's anti-fascist fervor was so quickly replaced by anti-communist fervor as the Cold War followed up World War II. But on the other hand, the revulsion to these totalitarian systems – labeled respectively right and left—is not so hard to understand. Both systems represented, to the majority of Americans, the extinguishing of hope.

In the animus toward Hitler and Nazism, there has never been a time when the American public was more unified in its hatred of a common enemy. No doubt every individual had his or her personal image of the threat represented by the dictator and the armies he led on a rampage of death and destruction, but one thing seemed obvious to all: in the path of Hitler's onslaught, human life became trivial, disposable, insignificant. He spearheaded a reign of desolation that gave no room for anyone to hope for a better, more peaceful life or a safer world. It was obvious that only by removing this scourge could his crushing influence be eliminated.

It is remarkable, then, that almost as soon as Hitler and Nazism were defeated, a new threat took its place—one that was almost diametrically opposite in its origins, purposes, and objectives. Taken at face value (which, it turned out, it never could be), Communism with a capital C represented a dangerous precept masked with a benign countenance. In theory, Communism presented the possibility of an unprecedented level of mutual accord and social equality that seemed custom-designed for industrialized economies. The fruits of labor would be shared by all. Property would be divided up, not according to the measures of wealth and acquisition, but according to what was best for society. All for one, and one for all would become the mantra of a social reformation that seemed destined to catch on around the world.

What became obvious—given the cruelty, paranoia, and rapacious power hunger of men like Stalin, Lenin, and Mao—was the harsh reality that hardcore Communism bore not the slightest resemblance to its theoretical justice and egalitarianism. As millions perished in Gulags and prison camps, and millions more were consigned to starvation through mind-boggling bureaucratic mismanagement, Communism easily replaced Nazism as the new ogre in the American psyche (not to mention, the target of enough warheads to wipe out continents). Those fiercely on the right or left may try to rationalize or justify the depredations of presumptively opposed systems of social engineering, but the leaders of totalitarian forms of government have clearly established—through the demonstrations of twentieth-century upheaval—that the evils of either or both systems vastly outweigh the benefits.

And something more. This trial-and-error period of social history has given Americans an almost truculent sense of righteousness when it comes to discussions of human liberty and individual rights. As anyone knows who turns on the news and reads the op-eds, we are wholeheartedly engaged in a sometimes raucous debate regarding the relationship between federal, state, and local government vis-à-vis big business, small business, families, and individuals. But nowhere, in this debate, do we find anyone seriously propounding (at one extreme) the takeover of a single dictatorial figure to manipulate society according to his own whims or (at the other extreme) the complete dissolution of government. We don't know where, exactly, we are between those extremes, and there's a good chance the border wars between individual freedoms and government inhibitions of those freedoms will remain perpetually in doubt. But through it all, and despite the loud yelling, the American public does seem to have some general agreement about who we are and what we want.

One thing we all want—I *think* we can agree to this—is fair opportunity. Certainly this concept is bandied about by all parties, on all sides, though the words "fair" and "opportunity" are loaded with multiple meanings. Is Opportunity, with a capital O, something that should be handed out in equal doses to every individual? Or is it something that needs to be earned or merited? If the American heroes are those who seize Opportunity, what can be said of those who refuse to offer it to others? And what of those who are incapable or unwilling to take the Opportunities that are offered to them? What of those who are blind or indifferent to Opportunity that is right in front of them? Is there a social and moral obligation to put some meaning into the phrase, "equal opportunity," or should we simply accept the fact that some people can grab the merest thread of opportunity and turn it into whole cloth while others can't even take the first step when the entire path is laid out before them?

Yet for all its ambiguity, and our uneasiness about who is "deserving" and who is not, we have reached a point in our cultural history where we can pretty much agree on what it means to *deny* Opportunity. We know what it was like during those centuries when African Americans were denied opportunity. We have taken enormous steps to offer women the opportunities once limited to men. And today we wrestle with our laws and our conscience over the issue of what Opportunities should be afforded to the most recent immigrants to these shores.

But somewhere between the ideal of "equal opportunity" and the abhorrent concept of denied opportunity, there is a reasonable middle ground which we Americans refer to conscientiously and repeatedly. It is the concept of "fair." In wage negotiations, in trade, in business, in government, fairness is the yardstick by which right and wrong, justice and injustice are measured. And in regard to Opportunity, as well, there is the unsettled and constantly debated issue of "what is fair?"

It's a worthy debate, because hope hinges on fairness. Whether it's the relatively far-fetched hope of winning the lottery or the more practical hope of getting an education, there needs to be some route to the destination that is not blocked by tyranny, corruption, fraud, or fear of reprisal. Those who seek Opportunity are not always going to be rewarded. Both willpower and waypower are subject to power failure. But despite success or failure, winning or losing, the strategy of hope is achievable if we all have respect for fairness.

But that, in turn, depends on trust.

Chapter 4

A Level Playing Field
Foundations of Trust

A car that's a racing machine is quite different from a car that's used by a family as a convenient a form of transportation. For different cars, different rules apply.

On a racetrack, speed is of the essence. Get your vehicle into peak condition, prepare its tires and supercharger and brakes for many circuits, put on your helmet, and get ready to go flat-out. Your intention is to pass and take the lead. Your methodology involves a none-too-delicate balance between preserving your life and testing the limits of your racing machine. To win, you will have to maintain speeds that would make your mother shriek. You will have to surge past other drivers who are just as intent on keeping you behind them, and you will have to make lightning-quick decisions when you have opportunities to pass. Your judgment calls, if erroneous, could result in maiming, death, and incineration.

However horrendous this may seem to a person of sound mind, the rules of the racetrack are expressly designed to encourage daredevil automotive performance. Obviously the rules of the track are not intended for Sunday-afternoon drivers cruising around with Mom, Dad, and the kids. The objective in race car driving is to maintain stiff competition—to pit skilled drivers, pumped up for this very event, against one another with the intention of determining who deserves first, second, and third place. There will be big winners.

There will be losers. There could be some disastrous events. But in this competitive environment, the rules are fair only insofar as they apply to a risky enterprise and to drivers keenly aware of the dangers involved.

For the rest of us who seek safe, efficient journeys on public roads, other rules apply. Speed limits are posted on road signs. We obey red lights, orange lights, green lights. Yield signs. Dashed lines (passing allowed) and solid lines (keep to your lane). And so on—not to mention the whole process of qualifying for a driver's license with a road and written test, registration and inspection of the car, and insurance requirements. Rules, rules, rules. None of them particularly onerous for people who hope to get from one place to another with their limbs intact and vehicles unimpaired—but rules that would be absolutely absurd for the competitive driver intent on outflanking, outwitting, and outperforming other competitive drivers on a track.

So—which rules are fair? Which satisfy the concept of a "level playing field?"

Different playing fields, different rules. As for which field is level, isn't it all in the eye of the beholder?

Social Acceptability

Before we leave behind the driving analogy and get to where we're heading (e.g., the world of stock portfolios, hedge funds, and credit default swaps), it might be instructive to look at one more traffic situation—one in which regulation, self-interest, mutual respect, free enterprise, and recognition of greater social good all play a simultaneous role. I refer, of course, to merging traffic.

You've been there. You know what it's like. Some lamebrain road engineers, in their infinite wisdom, decided that, on paper, it would be reasonable to bring in two lanes of traffic from the northeast and two lanes from the northwest and leave it to all those eager and anxious drivers to merge from four lanes into three (or eight into four, or five into two—we've seen it all). Anyway, here's what happens. Traffic slows, approaching this dreaded merge. In the soul of each individual driver there occurs a slight hiccup of consciousness as each one, individually, then collectively, begins braking for the inevitable. For some, the regulars, this is an expected delay, no more noteworthy than the rising or setting of the sun on a commuter's day. For others, it represents a shift of intended consequences, as they anticipate that arrival times must be changed, a meeting postponed, or a partner alerted. It is, therefore, with mixed emotions ranging from rage to resignation that drivers slow during their approach to this merge and begin to decide on individual actions.

Let's pause, right here, to note there are no real rules to this merger process. There are no regulations, for instance, saying that big vans have the right of way and Mini Coopers have to scurry off to the side. It is, in its outward manifestation at least, a dog-eat-dog, winner-take-all kind of merge. Whether you come through scathed or unscathed depends on your ingenuity and behavior of you and those of the drivers of the vehicles on all sides. It is, too, a situation where self-interest prevails and individual initiative is rewarded. However aggressive and pressed-for-time you may feel coming into this merge, there is absolutely no up-side to ramming another vehicle—a strategy that, you know damn well, would result in altercation, cops, insurance-company delays, and a lecture from your parent, spouse, or significant other. On the other hand, you'll gain nothing but honks and middle fingers if you hang back and decline to take advantage of every opening in the traffic.

In this state of heightened alertness and vehicular sensitivity, all approach the single, shared problem of squeezing traffic from many lanes into few. This is accomplished, most often, in nearly flawless ways. The shy and the belligerent; the fearful and bold; the hotshot in his Porsche and the grandpa in his Buick; fifty-foot trucks and nine-foot SmartCars; kids with brand-new licenses and limo drivers with millions of miles of experience—all, somehow, come to some grudging agreement about who precedes, who follows, who gets to edge in, who doesn't, who gets a warning beep and who slams on brakes in the nick of time. It's negotiation of a very beautiful kind—

exactly what we expect of ants in smooth collaboration, bees going about their hive-making business, and geese arranging their flight patterns.

How do we do it? Day after day, week after week, month after month? Why are there few collisions instead of many? Why does the merging traffic move forward over what is recognized as common ground rather than coming to a dead standstill over disputed road territory?

At Camp Wachusett we learned to work hard, play fair, and be a good guy.

It all happens because *we work it out.*

That's why.

And therein lies the real lesson about fair play.

Some Confusion about Acceptable Behavior

For years, the high-power business schools have been preparing kids for the competitive world of free enterprise as if it were a NASCAR race. Lessons, for the most part, are not focused on service to customers, contribution to society, or providing long-term support for Aunt Edna in her dotage. The working model of the economy, in the b-school view, comprises a host of self-interested individuals (and their respective enterprises) intent on creating value (read, profitability) for themselves and their stakeholders.

Viewed this way, there are winners and losers. Winners make big profits, build businesses, and leave their competition in the dust. Losers don't. They fail to achieve profitability, go out of business, and either disappear into the mists, flame-out, or prepare for a roaring comeback. The rules, in this model of the playing field, are much more akin to those of NASCAR races than those of merging traffic.

Nor is this view of business reserved to b-school culture. It is pervasive. Trophy-winning coaches turn into iconic figures of business mythology. Many guidebooks to business success emphasize swimming with the sharks, emulating ruthless warriors, and using the tactics of generals going into battle. Many a toast is offered to the hero who gets the better end of a deal, the tactician who performs a slam-dunk on competitors, or the financial wizard who discovers a loophole that no one had ever spotted before. All are signal victories for anyone on the fast track.

But what are the real assumptions behind this NASCAR version of business success?

Much depends on the way we perceive economic realities and, in particular, the nature and objectives of capitalist enterprises. Obviously if you're the CEO of a large corporation with numerous shareholders, you've got some important decisions to make when profitability sags and share price plummets. You can increase efficiency—lay off people, consolidate divisions, implement cost-saving measures. You can grow your market—figure out ways to expand clientele or beat out the competition. You can adopt new strategies—introduce new products or services that give you a fresh advantage. None of these actions are likely to make you feel warm and cuddly or put you first on the list of most-generous human beings. More likely, these profit-making strategies are going to involve some definitive measures and will be regarded as draconian.

Layoffs.

Early retirements.

Wage freezes.

Automation.

But, also, if all goes well, some of those tactics, if brilliantly executed, could help bring about turnarounds.

Efficiency.

Reliability.

Consumer appeal.

Market innovation.

When the goal is profitability, we take it for granted that it may take a sort of moral neutrality to meet your primary objective. Open-mindedness, generosity, empathy are all well and good. But they're not high-priority virtues when you need decision-makers who can close a division, implement production efficiencies, or come up with the next generation of products and services to feed unmet needs in the market.

Ultimately, though, even the toughest, seemingly heartless CEOs will tell you their surgical amputations and painful decisions preserved the company and enabled its survival. Their tough actions saved jobs for some employees and positioned the company to grow and resume hiring in just a few short years. Their actions led to a future instead of what appeared to be an inevitable funeral.

Profit-making business enterprise is not an activity that readily responds well to rules and regulations. As demonstrated time and again, where regulatory agencies try to *fine-tune* the strategic decisions of a profit-making entity, the results are likely to be laughable if not catastrophic for the corporation. It's all in the math. Let the costs of labor and raw materials outpace the price that customers are willing to pay for the product, and the enterprise will go belly-up. Cut off access to markets, or the means of serving customers, and disaffected potential purchasers will go elsewhere. Impose compliance regulations or taxes that result in costs that the market will not bear, and the business will collapse. In the profit-making world, there is only one math that ultimately means anything: income must exceed expenses.

And yet...

Within the mathematical boundaries that define a viable business, we recognize there are some approaches and tactics that fall within the range of acceptable norms of behavior and others that violate our sense of justice or earn our opprobrium. We're delighted with Apple's profitability because its products are works of wonder, and we're happy to seize the next iPod or iPad at a reasonable price. But on the other hand, rumors and reports of Apple's exploitation of Chinese workers make us uneasy. We don't mind (too much) the profitable oil companies that efficiently and steadily feed us the fuel we need to power-up our vehicles. But let there be an accidental rupture in a tanker ship, a tundra pipe, or a seabed drill, and we are filled with alarm about the reliability of the vast profit-making enterprises that find the oil, recover it, refine it, and deliver it to our internal combustion engines. Yes, profitability is a respected means to an end, but that doesn't mean it is sacred. The means are always subject to question, and the questioning is sometimes the measure of who we are and what we can accept as a society. This is an American value: we like low-cost energy and high profits, but we aren't willing to ruin our grandchildrens' planet to enjoy them.

We want to draw the line somewhere. But *where* we draw the line is the topic of endless fractious debate, and debate that tends to send us into opposite corners of the sparring ring. We can all agree, for instance, that buying and moving African slaves to this continent and putting them in the fields to pick cotton for the profits of plantation owners is shameful, immoral, and not something we want to see ever again in our society. But when immigrants slip over the border and go to work in the fields under conditions that would make a slave owner blush, we begin to get very uncomfortable with the discussion that follows. When we stretch our educated imaginations and try to envision what it would be like to work twelve hours a day in a pitch-dark mine occupying a crawlspace where it is not even possible to stand erect, we shudder. But when we're engaged to be wed and gazing on the gold and diamonds on display in the downtown jewelry store, we're not very good at stirring ourselves to ask where the gold rings and diamond settings came from, or how they got the price tags they sport under the gleaming glass counter.

The fact is, machines and institutions of profit deliver the goods, for which we have many reasons to be grateful. And these related discussions about where the goods came from, how humans were employed to obtain them, how the profits were made, and by what means these goods came within reach—all these considerations can quickly put us in a state of considerable discomfort. And, as is so often the case, when we are uncomfortable, the easiest and fastest thing to do is take sides.

Where Do You Stand on this Issue?

Ah, now we're at the core of left and right. (And already you're wondering which position the author is going to take!)

But I'm going to do something quite different here. I'd like to look at why we, you and I, so strongly feel the need to take a position when it comes to discussing what happens in the marketplace of human transactions—a market that has been evolving, let's face it, ever since we began sharing bison steak at the communal fire.

But before we get to the niceties of steak-sharing, consider our knee-jerk twenty-first century positions, left and right.

The left, in broad strokes, hates the exploitation of the workers and the excessive profits of the capitalist owners. The left would have a governing body, serving the will of the people and the ideals of a better society, step in whenever necessary to ensure that workers' rights are protected, that they are not exploited, and they have a broad array of social services ranging from guaranteed food and housing to guaranteed health care and education.

The right—again in broad strokes—will maintain that without the methods and means to achieve profitability, there can be no prosperity for anyone. Some people get richer than others because they work harder, have more ingenuity or initiative, or because they are luckier. Others may not be so fortunate, but that's just what happens in nature as well as society. Some make it, others don't. And history has proven that the most successful societies are those where free enterprise flourishes.

Have I adequately characterized—and caricatured—the hard left and the hard right?

I hope so, because, once entrenched in either way of thinking, we become highly motivated to find the proof that helps harden our attitudes. The true leftist will, unblinkingly, find evidence that workers have been exploited by capitalists throughout history, that the rich are highly motivated to ignore poverty and trample on human rights. At the extremes of the hard left, there is rage about injustice and a determination to right the wrongs that are perpetrated by the powerful elite. And there is entrenched resentment toward anyone with accumulated wealth.

Go to the fringes of the hard right, and you'll find equal and opposite rage directed at those who disrespect the very principles of a capitalist society; who claim equality as if it were some kind of God-given right, and threaten the very underpinnings of a free-market economy. Those entrenched in the most extreme positions of the hard right consider themselves assigned to mount a perpetual defense of property against those who would take it away. If the defenses crumble, anarchy will ensue.

Of course, on each side of the left-right fence, there are the great theorists. In the most conventional thinking, these would be Marx and his followers (left) and Adam Smith and his followers (right). Upon the foundation of these theoreticians, great bastions of economic and political thought have been built. And despite some notable and interesting crossovers, the economists, professors, writers, leaders, and pundits who started left have pretty much stayed in the left-hand lane while those who got launched on the right continued bearing to the right at full speed.

But recently, there's been some very interesting rethinking about these hardened positions.

The Brain on Trade-Offs

The ambivalence, as observed by a group of observers called neuroeonomists, is located in the human brain.

And it begins with that "uneasiness" I referred to earlier. As medical science, especially neurology, has become more sophisticated—thanks to the numerous kinds of brainscans and blood-chemistry analyses now available—and as research psychologists have begun to amass studies of neurological changes in human interactions, some very interesting findings have emerged. Broadly put, it is becoming apparent that most of us (not all, however!) have some kind of inbuilt, inbred, self-actuated, automatic tendency to understand, appreciate, and, if possible, practice a measure of trust in other people.

I don't think we have to look very far to see how this works. Just go to your corner antique dealer, pick out a charming little nightstand that you covet, and ask about it. Forget price for the moment. Let's start with the basics. Is it really an antique or a reproduction? Is it known where it came from? Are pieces like this rare or were they mass produced? Okay, now that's out of the way, let's get to price. A rare, hand-made, original piece—I'll pay good dollar for that. If it turns out I can't afford it, I'll sigh and walk away and hope that Uncle Bennie leaves me a small fortune, all without any negative feeling toward the antique dealer whose goods I admire.

But what if this attractive nightstand is a knock-off reproduction, albeit with sturdy legs and good utility value? It's still just right for my bedside table. As long as I know what I'm getting, I may take it anyway. But I'm prepared to pay a lot less.

Notice that, once the basics have been established, you and the antiques dealer can do a lot of haggling over the actual price without anyone feeling too offended. What *is* offensive—supremely so—is the dealer who sees that you don't know much about antique nightstands and goes about convincing you that a reproduction is an original.

As neuroeconomists look at situations like this, they see a couple of elements come into play that will actually have a big influence over the transaction. One is social. The other is chemical.

From a social point of view, it makes a big difference whether you're a regular at the store or a one-time visitor. The deal you arrived at—and the way it's cut—will probably be quite different if the dealer knows you live right around the corner rather than being a tourist in town for the weekend. This may seem self-evident, but the deal is likely to be a lot more fair and equitable if the antiques guy thinks he'll see you in the neighborhood and have to look you in the eye or if he thinks you might come back soon to purchase a matching dresser. This is about consequences: painful or pleasurable.

So that's the social part. The chemical part is more difficult to detect, but also quite powerful as an influence. This whole transaction with the antiques dealer is not just one computer doing calculations with another. While you're talking about the provenance of that nightstand and listening to the dealer's claims and checking out what he's wearing and the note of sincerity in his voice, your body is producing—among other things—a chemical called oxytocin. Exactly why, and how, is now being much-studied by researchers like Paul J. Zak, author of *The Moral Molecule*. But what we do know is that this chemical makes you feel good about what's going on. Depending on *his* body chemistry, that antiques dealer may or may not be getting a similar kind of rush. But if he is, then he too is feeling pretty good about what looks—from the outside—like nothing more than a commercial transactions. That feel-good chemical is going to affect both of you during your current transaction. It will also influence the way you approach each other in the future.

Look what just happened. In the process of this mundane economic transaction—I want the nightstand, he wants the profit—we have stumbled into a wide assortment of influences that have nothing whatever to do with the authenticity and worth of the table. These other factors are what we call (to the great discomfort of traditional economists) *feelings*. And what neuroeconomists are exploring, in new depth, is how these feelings affect not just haggling with antique dealers but the whole realm of economies.

Can I trust the person I'm dealing with? Is he lying to me? Did he murder an old pawnbroker to get this piece that he's trying to sell me? Did he pay pennies on the dollar to some wage-slave to paste together a reproduction that he wants to sell at 10,000 percent profit? Will this be the start of a twenty-year relationship during which he learns my tastes, calls me up when he's got something I might like, and helps me furnish my entire home? Or will he be satisfied never seeing me or hearing from me again?

All these questions, and more, are going to influence our future economic transactions. (And, just as an aside, note the distance between this transaction and any kind of doctrine. Even if I am on the far right and believe this dealer is deserves of every penny of profit he can squeeze out of me, I will be incredibly pissed off if he sells me what's supposed to be a Queen Anne table and it turns out to be a reproduction. Even if I am on the far left and believe that the dealer is a ruthless capitalist trying to rip off the common folk like me, I would be an idiot to "liberate" his property just because I happen to like the looks of it.) In any such transactions, we rely on some degree of fairness and some measure of trust. Where fairness and trust are lacking, no matter what our political persuasions, we get uneasy.

Yes, our society has race-car drivers. But on a crowded highway with merging lanes, they're not the ones we trust. And it's trust that makes it possible for us to merge safely without creating a demolition derby every time we negotiate private and public space.

What Happens When Trust Falters? (A First-Person Account)

Chris Tabor (as I will call him here) wanted investment advice. No problem. That's what I'm here for. Where should he put the funds he was going to draw down in retirement? What were some good ideas for high-risk, high-return? What were the characteristics of companies I believed in? What did I think was going to happen to Treasuries in the future? What about real estate as an investment?

So far, so good. Usual questions. Not easy to answer without getting a more thorough understanding of his assets and priorities. But I generally got an idea of what he wanted and how his funds might be allocated.

Then his questions took off in a different direction. He wanted to know which bank I liked to deal with. Could he talk to some clients who had been with me for ten years? Someone else who had stayed with my firm for twenty years? Could he get twenty-four-hour information about the allocation of his funds? Who was our accounting firm? Where could he get audited reports on my firm's performance? Had we ever received a disciplinary action from the SEC or other law enforcement agency? Had we received letters of complaint? Had I been doing business with any hedge funds? Which ones? Why?

And on...and on...and on.

Mind you, all perfectly good and legitimate questions. Many were the kinds of questions you *should* ask before placing a significant part of your wealth in the hands of an asset manager. But still, for me, unusual. Prospective clients are usually optimistic and hoping to find a fit, to find ways to form a mutual bond. Most of my new clients have previously seen me on TV or heard about me or Farr, Miller and Washington from other clients. They wouldn't bother picking up the phone if they didn't have reason to trust our firm, based on what they'd heard from friends or colleagues. But Chris was on full-alert aggressive and sounding fearful. As I fielded his questions, I got the sense that nothing I could say or do would be enough to address his uncertainty and meet his concerns.

So I asked him: "Chris, what's been your experience in the past?"

He looked chagrined. And his answer came quickly: "I got screwed."

That's when I got the full story. He wasn't kidding. In the steep run-up of the market between 2001 and 2007, he'd been wooed and won-over by a guy who could have been a dead ringer for Bernie Madoff. Smaller scale, but same-style Ponzi scheme, where the guy constantly gets new investors so he can give previous investors a nice-looking payout. It was the kind of scheme that only works as long as the market is going up. Which, of course, it suddenly stopped doing in 2008. Which is when Chris Tabor got burned. Seriously.

So...what could I tell him? "Chris...trust me!" Ah yes—the words of every slick salesman and con artist. The very words, no doubt, spoken by the Ponzi schemer who had just snatched away thirty percent of Chris's life savings. What I had to tell Chris was devastating for me. I told Chris that while we had no letters of complaint or any disciplinary action from the SEC or other body, and while he could investigate and talk with clients and others, there would be no way for him to be sure that he was safe from being ripped-off again. No matter what prophylactic steps one takes, there will always be a way for a criminal to do you harm. Therefore if one's well-being is always in part illusion, then all relationships rely on a foundation of trust.

But what *would* it take to restore his trust? How many questions would need to be answered before he could be satisfied that someone was truly looking out for his interests and treating his money with respect? How many reports and audits would he need to help him recover from the blow he had been dealt? And how long and elaborate would this whole process be—the most challenging task that anyone in business can face—the job of *restoring trust?*

I was about to find out.

As I write, a number of years have passed since I took on Chris as a client. He's great. We get along fine. I think he's been satisfied with what's happening with his investments, the way we report, and how we behave as a company. But that other element—trust? Is it back? How long will it take before that's *really* restored? If ever?

It makes you think. Chris is just one guy who got burned. But what happens when there are thousands like him? Millions? When a whole society and economic system goes through a phase when lies abound, promises are broken, clients are cheated, chicanery is accepted, and fraud is unchecked? Back in the old days of hellfire and brimstone, those who committed such offenses against their fellow humans would have been cast to the lowest levels of Dante's Inferno, and we could revel in poetic accounts of their suffering. Not today. These days, the clean-up crew is right here in the trenches. It's all of us who face the arduous task of trying to restore trust that has been broken. Which is about as simple as trying to restore faith in humankind. Not easy.

Today, I see an entire economy experiencing a crisis of trust. By "crisis," I mean things can go either way. For sure, there are still huge reservoirs of trust in this society, and I am grateful that an immense number of transactions take place without the taint of graft or even tomfoolery. But I also see some large gaps opening up—particularly during a financial crisis like the one we have been through—when people like Chris Tabor come out thinking, "Who am I dealing with?", "How can I trust them?" and "What am I setting myself up for next?"

The consequences of a trust crisis are serious. In 2009, hard on the heels of the near-total collapse of the U.S. banking system, there was a dramatic shift in the public's attitude toward the banks and the stock market. According to a BBB/Gallup poll conducted in April 2008 (before the financial crisis), 42 percent of Americans said they trusted financial institutions and 34 percent said they trusted U.S. companies. Those figures, themselves, are nothing to celebrate. But as the financial system was shaken by Wall Street's earthquake, the confidence in financial institutions plummeted to 34 percent and the trust in U.S. companies fell—within a year—to a meager 12 percent. As for the public's view of our banking system: the Chicago Booth/Kellogg School Financial Trust Index reported that between September and December of 2008, 52 percent of Americans simply gave up having any trust at all in banks.

A temporary setback in the public mood? A little stumble in confidence from which we, as intrepidly optimistic Americans, will soon recover? Well, let's hope so. Because trust is a valuable and important component of a well-functioning economy. Steve Knack, a senior economist at the World Bank, has been studying the "economics of trust" for more than a decade. His conclusion? "If you take a broad enough definition of trust, then it would explain basically all the difference between the per capita income of the United States and Somalia."

Can we put a dollar figure on that? Do we need to?

Better, I think, to simply figure out a way to hold on to trust, by the means I will propose.

It is an incredibly valuable asset. And it's in crisis.

Chapter 5

When Trust Erodes
Breaking Our Promises

Here's the experiment.

We (the experimenters) are going to give you (the participant) the opportunity to make few bucks. The rules are thus:

We will give you $10 to begin with. That's yours—just for showing up.

But we'll also give you the chance to make a little more by working with another participant you will never see and never meet. Here's how: The unseen participant *also* has a small bankroll, and she is at liberty to send some of that money to you. As soon as she does, and when the money hits your account, the amount that has been sent to you will *triple* in value. In other words, if that other person sends you $4 from her account, you'll actually receive $12.

So that's a nice windfall—right? You started with $10 and because of some other person's generosity you could walk away with $22.

But we're going to give you the opportunity to take one more step. Rather than take your windfall and go home (which you're perfectly free to do), you also have the option of sending back some of your earnings to that unseen benefactor. How much? Well, that's up to you. You might figure, well, she had $10, she spent $4, so she's got $6 left. If I give back just half of my newfound wealth, $6, she will come out of this with $12 in total and I'll have $16. It's a win for both of us (though I win more). On the other hand, I could just give her back $4. That would make her whole, and I get to keep $8, so she gets her original $10 and I get $18. On the other *other* hand, why not just keep it all? When she spent the four dollars, she had no idea how much I would give back—if anything. There were no conditions set. And since all the participants in this experiment are kept anonymous, she'll never know I kept her money, and I'll never meet the person I accepted it from.

Needless to say, this isn't a game I thought up in my spare time. Rather, it's a very carefully devised experiment that has been replicated many times, with different groups of people and variable amounts of currency, by psychologists and economists who are trying to get a read-out on the way we humans behave when presented with certain economic situations. It's called, for short, The Trust Game.

And the results?

According to Paul J. Zak, a neuroeconomist who has repeatedly tried the experiment with vastly different groups of people, "With large sums or small, in dollars or dinars, participants almost always behave with more trust and trustworthiness than the established theories predict they will." In his own experiments, Zak reports, about nine out of ten participants would take the risk of "investing" their money in some other, unknown person (like you)—and nearly 95 percent of the time, you would send some money back to your unseen benefactor.

And the question Zak asks, of course, is *why?*

Good question.

If you went to business school, or just took an economics course at some point in your education, you probably learned the assumption about economics attributed to one of its revered founding fathers, Adam Smith. Essentially, according to this perceived wisdom, it can be predicted that each of us will act out of self-interest. The self-interested individual (academicians actually dub him *homo economicus)* is supposed to be wondrously and consistently self-centered when it comes to the accumulation of wealth and property. As Zak points out, in the view of "traditional" economists, when there is money to be made, *homo economicus* will try to make the biggest bundle possible. When there's property to be acquired, *homo economicus* will acquire like crazy. From the hard-headed perspective of traditional economists, it's acquisitiveness that makes the world goes round. Or, as Zak puts it, "Economists have fallen in love with a concept called 'rational self-interest,' which assumes that each individual makes decisions on the basis of personal advantage, and also on the basis of a rational calculation as to exactly where that advantage lies."

My take on this is a bit different. Keeping in mind our thesis of "affordable morality," let's recognize that each participant is participating in a game or study and has been provided an initial stake. There is no desperation to survive. Therefore it's not that hard-edged economic self-interest is wrong, but that it will harden or soften behaviorally as the real economic condition of the participant hardens or softens.

Years ago at Pomfret School, my colleague Father Jim Birdsall gave a sermon on Jesus' miracle of multiplying the loaves and fishes. Jim suggested that the miracle may have been one of human faith and generosity. Jim's idea was that the majority of folks in the crowd had food with them, but seeing no food in the open and so many people, decided to keep their food hidden. Jim suggested that perhaps the real miracle was Jesus' first sharing of his seven loaves and fishes that inspired the rest of the crowd to reach into their cloaks and share theirs.

In Jim Birdsall's explanation, as desperation was reduced, generosity increased, trust increased, and the society benefitted.

In hindsight, it's remarkable that the concept of *homo economicus*—a strictly one-sided approach to the understanding of motives—has held on so long. But such a theory had two advantages. It made the "formulas" used for economic analysis and prediction much more manageable. (What economist in his right mind would want to try to factor in the array of emotions—ranging from outright greed to pure generosity—that might compel human behavior in wildly unpredictable directions?) And for those who made profit their religion, this formula provided a huge degree of moral justification. ("Of *course* I want to rake it in this quarter—I'm a businessman, not a nun.")

But when experiments like The Trust Game put the whole notion of "rational self-interest" to the test, the motives and rationale for our actions turn out to be a lot fuzzier than they would be if we were actually just self-acquisitive automatons. To see why, just consider what you actually *would* do if you were one of the participants.

Suppose your unknown benefactor gave away her entire wealth ($10) to you. If your self interest is finely-tuned and well-honed, you will gleefully walk away with $40 (the $10 you started with, plus the $10-tripled that your benefactor gave you). Presumably she'll walk away with nothing, but so what? This isn't a game of morality. It's a game of money. And if she gave away all that money, she loses, you win.

But somewhere, deep inside, you may hear a little voice saying, "Wait a minute. Is that fair?" Is *what* fair? What *is* fairness? Well, when your benefactor emptied her bank account, don't you think she may have been thinking that you would both be winners? Give her back twenty and you've still doubled your money. She must have been trusting you to do that. Or was she? Who knows *what* she was thinking?

Okay, but what if she only gave you two dollars—tripled, that means you got six. So she only trusted you a little bit. Should you give her a big reward if she gives you a big dose of trust, and a little reward if she just trusts you a little?

And all these questions get even more complicated if you subscribe to, or are influenced by, some set of moral or ethical principles that interfere with your rational self-interest throughout this whole transaction. What if, for instance, you have preconceived ideas about generosity—and you have an unquenchable desire to give back some of the money *just because it feels good to do that*? (What are you—nuts?) Or you have some notion about justice (divine or otherwise), and you really think it's just not right to get a windfall without sharing it with someone. Or, even stranger, you are able to imagine the disappointment of your unknown donor, and you can further imagine the feelings of guilt that you'll have if you don't do something to reward her generosity.

A game it may be. But all the conflicting thoughts, emotions, and desires that enter into this game are quite messy. And the amount of trust that ends up being shared between you and the other participant—well, that all depends on who you are and what you consider to be important.

A Further Twist

In the version of the Trust Game that I have just described, none of the participants knew anything about anyone else who was playing the game—and it was done in a setting where it was impossible to find out (either during or afterward) who has been trusted or trusting, or that is, who had been generous and who had not. In another version of this experiment, three Wharton professors added other elements to the design in order to address some of the more complicated issues having to do with trust.

What happens, they wondered, if the other participant is blatantly untrustworthy? Is trust so fragile that, once broken, it cannot be restored? And what happens if there's outright deception? How long does it take to mend *that* fence?

In the experiment designed by Maurice E. Schweitzer and John C. Hershey (both professors of operations and information management) and Eric T. Bradlow (professor of marketing), there were two sets of participants who were asked to deal with each other over a total of six rounds. In some ways, this experiment resembled the Trust Game just described, but with a number of crucial differences.

Let's say you were the innocent subject. As in the previous experiment, you'll be given some money to start. But this time, there will be six rounds, and on *each round* you'll receive $6. Out of sight somewhere is the "partner" you will never meet. You have been told that you have two choices. You can keep the $6 for yourself. Or you can send the full sum of $6 to your unseen "partner"—and if you do, the money will triple in value, which your partner can share with you.

Just before Round 1, you get a message that says, "If you pass to me, I'll return $12 to you." So you send $6 to your partner, but…too bad, you get nothing in return.

However, you do get a second note from your partner that says, "Let's cooperate. I'll really return the $12 this time."

OK, now what do you do? You've been deceived once, but at least there's been some communication—so someone must be there, a real person, and they've renewed their promise.

So, on Round 2, you send your partner another $6.

Again, you get nothing in return.

Twice, you risked your money. Twice, you were deceived.

And there are four rounds still to go. What will you do?

The question the researchers are asking (though you don't know this, being an innocent subject) is *what will it take to restore trust?*

To answer this question, the Wharton School professors set it up so that in the next four rounds every person who continued to risk $6 would receive $12 in return. The only difference was, half of the subjects (131 out of 262 participants) *did not receive any notes* before the first two rounds. So, for half the participants there was just untrustworthy behavior—no promises made, but nothing in return, either. For the other half there was the experience of outright deception—someone promising to do something and not delivering.

They discovered that it was much more difficult to restore trust after there had been deceptive behavior. For those who received no notes—therefore, no outright deception—trust was restored surprisingly quickly as they began to receive money from their partners on the third, fourth, fifth, and sixth rounds. But if someone makes a promise, deceives you, makes another promise, and deceives you again, trust gets blown away, no matter how your "partner" behaves in the future.

In What Do We Trust?

So—what has happened to the level of trust in America during the past couple of decades? Do you trust Wall Street, your Government, big banks or big business? And if you trust them less, how have your willingness to work, save, invest and spend been affected? Are you willing to trust other participants with a portion of your $10 (or $10,000 or $100,000) with the expectation of getting something more in return? Do you feel as if trust has been broken in some way? Or, worse (for the restoration of trust), do you feel that you have been the victim of some deliberate deceptions?

It's important to answer these questions because trust is, in so many ways, the glue that holds society together. This may seem obvious, but consider the benefits that accrue from trustworthy behavior.

Lower transaction costs. Consider the savings in a handshake agreement. You settle on a price. You settle on a delivery date. You shake hands. If there's true trust in this agreement, there are enough *unspoken* codicils and footnotes and provisos and wherefores and hereunders to fill several megabytes of documentation. If there is trust, it means that if the delivery doesn't work out, we'll discuss it and work out different arrangements. If the product is faulty or the service isn't what we agreed to, we'll either adjust the price or make some arrangement to fix what's wrong. If either of us gets hit by a bus or struck by lightning, the deal is off. And so on.

Now imagine if you have to get that handshake deal into a legal document. Well, you don't have to imagine. You know what those documents look like. Pages and pages and pages covering every possible situation that could arise, and then at the end of the document, an agreement to arbitrate or litigate or in some way protract the negotiations into infinity. The less the trust, the longer and more detailed the contract has to be.

And your legal fees (in a trustless environment) are just one of the transaction costs. How about the inspections that must take place?

How about the review process? How about hiring the people who have to get signatures and give final approval on everything?

Of course, most of us operate somewhere between these two extremes, with a handshake signifying an initial agreement on the assumption that details will be spelled out later. When disagreements arise, some get settled quickly while others blossom into litigious free-for-alls. But the point is, where trust is lacking, every additional degree of wariness and oversight and monitoring costs time and money and resources—escalated costs that must be born by the participants.

Less policing. Some of our grandparents can still remember neighborhoods where it was safe to leave the door unlocked. No crime, high trust. Now, of course, we live somewhere on the continuum from locked doors and neighborhood watches to a cop on every corner and double-bolted doors. And that's just in our neighborhoods. As you remove your belt and shoes and empty your pockets at the airport, think what it would be like to live in a world where every airline passenger could be trusted. When you pay your next fine for exceeding the time limit on a parking meter, consider a fantasy world where no parking enforcement officer would be required because everyone could be counted on to obey the parking regulations.

Whether trust is shattered by monumental acts of terrorism or by trivial neglect of feeding the meters, count on it, there will be an associated cost to society in the form of the policing that's required to oversee, adjust, and control our behavior. In fact, to the degree that trust begins to break down, increased policing of all kinds may be required to maintain the semblance of social order. Watchdogs and regulators will be sorely needed. Enforcers will have to be on call. All that policing has associated costs in money and human resources, but even so, as trust falters, something or someone has to take measures to enforce rules, laws, or regulations.

Comfort with debt. Let's try two scenarios. In the first, a guy you've never seen before stops on the street and asks you to loan him sixty bucks. That's the word he uses—"*Loan!*" When you stop laughing and realize the guy is serious, you say, "Well, let me consider it. When will you pay me back?" "Next week," says the guy. Do you hand him sixty bucks?

Let me guess. Probably not. But why? A, because you've never seen the guy before. B, because you have no way of finding out whether he's really in the habit of paying back money that is "loaned" to him. C, because he doesn't even give you a clue what he needs the loan for. Cigs? Booze? A charitable contribution to the Boys and Girls Club of America?

Scenario two. Your cousin Tom, whom you've known since you were three years old, asks to borrow sixty bucks. He says he needs it for gas money. He's got a job that he'll finish up next week, when he gets paid, and that's when he'll pay you back. Do you hand him sixty bucks?

Well, unless Cousin Tom is the family's most notorious deadbeat, you'll probably give him a few twenties. Why? A, you've known the guy forever. B, you know where to find him if you need to. And, C, giving him gas money is a credible reason for making the loan.

But what's the real difference between the guy on the street who hits you up for a loan and your Cousin Tom?

Trust is what makes debt work. Whether you extend credit out of the goodness of your heart (let's help out Cousin Tom) or because you want something in return (let's put our money to work), you have to be confident that you'll be repaid, with or without interest. Or—and this is an interesting aspect of debt—you need to have confidence that your contribution will be used in the way that your debtor says it will be used. (Consider how many supporters of good causes, or contributors to charity, refer to their donations as "paying a debt to society.") But whatever your reason for loaning out money, the oil that lubricates the beneficent cycle of payment and repayment is trust.

In a capitalist society where incurring and repaying debt is a force, there are numerous ways to establish trust. The credit-checks that shadow every one of us through our financial lives. The family, professional, and personal relationships that enable us to build trust through openness, proven reliability, and fair dealing. The professional organizations and the legal oversights that help to enable others to trust us. None of these assurances is a rock-solid guarantee that we'll pay back our debts, but each connection is another bolt or strut in the scaffold that supports our trust.

And debt is powerful because it has forward momentum. By incurring debt you can build a factory, expand a franchise, or launch a start-up far more quickly and ambitiously than would be possible if you had to scrimp and save to come up with the investment capital. With debt, buildings go up faster, inventions come to market more quickly, and resources are put to use more efficiently. But, note, none of this is possible in a society where people do not have the ways of establishing the basic levels of trust that are necessary to assure that promises will be kept, debts will be repaid, and funds will be put to the use for which they are designated. Take the bolts or struts out of this structure, and the whole edifice of trust tumbles down.

Institutional stability. Hardened veterans of dysfunctional companies know all-too-well what it's like when Sid, the vice president of sales and marketing, sets out to undermine Mary, the vice president of operations. Each is suspicious of the other. They stop talking, and the direct reports on Sid's team devote much of their time to proving that Mary's team is incompetent, while Mary's team, in turn, sets up a defensive bulwark around their operations. Accusations fly thick and fast; cooperation ceases; the customer suffers and so, eventually, does the company.

Institutions depend upon a certain basic level of operational trust for their survival. Somewhere along the line, the incentives for cooperative action (common good) have to outweigh the motives of pure, exclusive, individual self-interest (personal gain). Simply put, Sid and Mary have to work together. And for that to happen, there must be some level of trust between them.

Ability to plan for the future. Once upon a time, visionary concepts of social engineering were held by many in high esteem.

By now, it's apparent that attempts to make societies work like elegant pieces of machinery are doomed to failure. Yet all of us who are engaged in this process of making things better for ourselves and our children would like to do some tinkering—either major or minor—to come up with some improvements. And the ability to make that happen, in any society, seems to depend largely on the level of trust that exists.

Trust, as we know, can ebb and flow, and every society is caught somewhere in the tide. Convulsed by war, crime, revolution, in conditions where trust is nonexistent or nearly so, there is no future. There is only the immediate need to defend, protect, and survive. At the other end of the spectrum, in societies where high levels of trust prevail, even when there is no perfect harmony (there can never be), no precise social engineering (it's impossible), there is, to lean on the cliché, the prospect of a brighter future. And that is the very essence of hope.

Trajectories of Trust

I ask again: What has happened to the over-all level of trust among Americans?

That may seem like an abstract question, but in recent years quite a bit of study has been devoted to the subject.

In his book *Trust: The Social Virtues and the Creation of Prosperity* published in 1995, economist Francis Fukuyama observed, "One of the most important lessons we can learn from an examination of economic life is that a nation's well-being, as well as its ability to compete, is conditioned by a single, pervasive cultural characteristic: the level of trust inherent in the society." In the years since, comparative studies of prosperity and levels of trust in various countries have gone a long way to support Fukuyama's observation. In a summary of research published ten years later in *Nature* by economists and psychologists in Germany, Switzerland, and the United States, the authors begin by stating, "Trust is indispensable in friendship, love, families and organizations, and plays a key role in economics and politics. In the absence of trust among trading partners, market transactions break down. In the absence of trust in a country's institutions and leaders, political legitimacy breaks down. Much recent evidence indicates that trust contributes to economic, political and social success."

But there is also considerable evidence that the United States, particularly in recent years, has seriously fallen in rank from being a high-trust society to becoming a nation where trust—in its many different forms—is seriously in jeopardy. This is particularly true of people's trust in financial and political institutions. A Gallup poll, showing the loss of faith in institutions between 2002 and 2010, puts some figures on this phenomenon. During those eight years, people's trust in banks fell 24 percent. Trust in the media plummeted 14 percent and confidence in U.S. public schools declined nearly 10 percent. (As for confidence in Congress—which comes in dead-last among ratings of sixteen public institutions—the Gallup poll figures fell from around 30 percent to 11 percent over a similar period.)

Many explanations have been offered for the deterioration of trust that seems to imperil our prosperity. When it comes to the collapse of confidence in financial institutions, we need look no further than the debacle of 2008 when the largest banks and insurance companies were unable to scramble back from the brink of collapse without hefty donations from public coffers. Small wonder that the ensuing years have not brought about a restoration of confidence in these institutions. As noted, obvious malefactors were slow to be punished—if at all. The trading practices that clearly contributed to the banks' downfall continued with new vigor—even when it became apparent that banks deemed "too big to fail" were growing even bigger. Regulatory bodies such as the Securities and Exchange Commission, the Office of the Comptroller of the Currency, and The Financial Industry Regulatory Authority (Wall Street's "self-regulator") continued to display their impotence when it came to protecting consumers from catastrophic loss. And congressional attempts to expand oversight have proven to be unwieldy and unenforceable.

On the political front, the ebb of trust has grown ever swifter with shocking stalemates that, in effect, have almost brought government to a standstill. The juggernaut of an expanding deficit; the inability to make choices that serve the public interest; and the cynical acceptance of the role of money in politics have all contributed to a public feeling that government simply cannot be trusted. And these public demonstrations of distrust affect every aspect of our daily lives.

Jobs and benefits and dreams. How many young people, today, are entering the job market with the expectation they'll have lifelong employment in one company? Or even one industry? Or that the job for which they've prepared will even exist in the next ten or twenty years? (Even in Japan—where loyalty was once rewarded by lifelong employment—there has been a cultural shake-up as once-reliable companies have shed their most devoted salarymen.) In the U.S., of course, the mania for acquiring and merging, downsizing, automating, and streamlining has been accelerated by a craving for productivity and, perhaps most of all, by the pace of technological change. Outsourcing is endemic. Loyalty to an organization, while still widely lauded, is temporary. The writing on the wall says quite clearly, "No employee is indispensable." And just as indelibly, there's writing on the opposite wall saying, "No company lasts forever." If hugely "successful" organizations like Enron and Lehman Brothers can go down the tubes, if Chrysler and General Motors can teeter on the brink, if the nation's largest airlines can go bankrupt, today's employee would have to be wearing blinders and earplugs to believe in the chimera of guaranteed lifetime employment.

The aforementioned B.F. Skinner studied the behavior of rats that were rewarded or punished to modify their behavior. His work provided great insights into pleasure pursuit and pain avoidance. When young people enter the workforce, they have the benefit of not only strong, vibrant minds and bodies, but also of their dreams, desires and limitless possibilities. After all, the U.S. is the country where any child can grow up to be President.

The pursuit of pleasure and avoidance of pain drive many careers. The dreams of wonderful possibilities inspire determined hard work and lots of achievements. But what if young people believe that possibilities and dreams have limits? What if, as in the Soviet Union, our emerging workforce was told they won't ever be allowed to own a home or accumulate personal wealth? How hard do you think they will work? How will they feel about their careers?

So, too, with the "benefit packages" that once seemed reliable. Healthcare coverage is a moving target, negotiable year by year, with employers and employees both struggling to deal with exorbitant costs that put both parties on the defensive and make the future seem very uncertain. Retirement plans that used to be just that—*plans*—have turned into improvisational pastiches comprised of 401k's that require deft distribution and Solomon-like wisdom to manage. Bonuses and stock options are used as powerful employment lures, but in highly dynamic industries where conditions can change with the speed of an electron crossing a circuit, those enticements will mean nothing if a start-up founders or a darling of investors turns into the unwanted hag of yesteryear.

A dynamic market, yes. Very high productivity (nothing to scorn on that score). But trust in tomorrow? A feeling of confidence and certainty about one's own job and career and future? Not featured in the current climate.

Investments. What about trust in the financial markets? For most Americans their homes are their biggest investments. Or were. Or were *once*, then *weren't*, and now *might be* again, if...

And in that "if" is a whole realm of possibilities. *If* the neighborhood around us does not get riddled with foreclosures. *If* there's a turnaround in the glut of houses for sale, so once-valuable property is again in demand. *If* the job market recovers and confidence is restored so that people now renting will begin buying again. *If* we are not swept, once again, by the kind of financial-market panic that hacked us to pieces in 2008. Those are a lot of *if's*—and every one of those *if's* signifies a decline from the sure-footedness that homeowners felt just a few years ago when the home was not only a home but an investment we could count on, borrow against, plan to leave to our children as a significant inheritance, or sell off to cushion our twilight years. That, for sure, is something you can't trust in any more—no matter how sizable your home, where it's located, or what balance remains on your mortgage.

So much for the average American's largest asset. What about all the other investment opportunities that we used to be able to classify, so sagely, as "high risk," "medium risk," or "low risk?" To begin at the bottom of the totem pole, the once highly secure and reliable low-risk investments don't even qualify as investments any more. A T-bill that will pay out less than the rate of inflation? A tax-free municipal bond with returns in the low single digits? These are not investments that will guarantee you a sustainable lifestyle upon retirement.

Well, what about *high* risk then? Yes, with enough capital and an iron stomach, you could think about playing with the big boys. Get into some of those stupendous pot-at-the-end-of-the-rainbow hedge funds. Maybe even get into currency exchange (if your stomach is titanium) as the Euro goes flirting with its own demise and capital flows move into other currencies.

If you're not up for the low-risk, no-return alternative, or the high-risk, head-on-the-chopping-block choice, what about the middle road—the bastion of secure-looking blue-chip stocks that you always felt you could…trust? Yes, it's still possible. But growth and profitability have to be very carefully monitored. The same acceleration of technological development, opening of world markets, and responsiveness to shareholders that has so markedly altered the job market has also affected the permanence and durability of corporations that used to be called blue chip and companies that used to be regarded as solid. Keen vigilance is required if you don't want to find yourself holding shares of TWA, Kodak, Sears, or U.S. Steel in their waning days.

Education. Now here's something you could always trust, perhaps the most valued asset in American enterprise. The guarantee of public education became a stabilizing force throughout the growth of the republic. For those who achieved and excelled and had ambition, steps through higher education would lead to upper plateaus of professionalism, guaranteed employment, comfortable salaries.

And now? Who can trust that a young person with a high school diploma has basic language and math skills? For the achievers who go on to two-year or four-year colleges, will their employment opportunities upon graduation allow them to pay back their student loans before they're in their forties or fifties? For those getting professional degrees, will their skills still be in demand—for ample compensation—by the time they graduate? Or are many with educational ambitions taking a risk that resembles the high-stakes gamble of a hedge-fund manager?

Participation in government. In the 2008 presidential election—one with the highest turnout in decades—36 percent of Americans did not vote. (There were even more nonvoters in the 2012 race.) In the 2011 local and state elections, fewer than 50 percent of registered voters showed up at the polls despite the fact that many of these elections would decide governors, state legislatures, as well as budgets and policies having direct impact on local communities.

Of course the reasons for *not* voting are as varied as the number of nonvoters. Some pundits would chalk it up to the large catch-all category of "voter apathy." Others report the technical and logistical difficulties of registering to vote and getting to the polls. Lack of knowledge about the candidates and a dearth of interest in local and regional politics undoubtedly make some people feel too uninformed to cast a meaningful ballot (particularly in local elections). And, yes, there are large doses of cynicism ("What does my vote matter? They're all crooks!") as well as outright hostility toward governance ("the best government is no government" philosophy). For the extreme view on voluntary disenfranchisement see P.J. O'Rourke's very funny book called *Don't Vote! It Just Encourages the Bastards*.

Many of us and many of our next-door neighbors just aren't taking part in the baseline process of government. And what that means, to all of us, is that it erodes our trust in the outcomes. Those 36 percent who couldn't or didn't vote for a president of the United States—what's bothering them? What are their needs, their wishes, their interests, their hopes and desires? It's wonderful that 64 percent of us, in 2008, made it to the polls where we had a choice among candidates. It's even more wonderful that we did not have guards with rifles peering over our shoulders while we cast our votes, or officials in unmarked vehicles shredding our ballots after they were cast. But, still…can we really trust election results that only reflect the views of two out of three Americans?

At the state and local level, the lack of participation is even more significant. Why so little interest in who becomes mayor? How can we ignore the people going into office who will be looking after our schools, roads, playgrounds, police, and fire stations? Again, the reasons for nonparticipation range from apathy and indifference to cynicism and disassociation. But the net result is that we have a mysteriously composed quorum of voters—usually a minority—playing a big role in determining the quality of our daily lives. And, having put our governance in the hands of this minority, do we trust their decisions?

It's no wonder that trust in government is in the low double-digits. Most of us, at some level, have simply ceased to be participants in the governing process.

A Restoration Project

So…what will it take to restore trust in our businesses, our financial system, our schools, our government?

Perhaps it's instructive to go back to the Trust Game I described at the beginning of this chapter. Remember that $10 we gave you—with the option of keeping it or "investing" some portion of it with an unseen participant? Well, let's re-label that $10 and call it "your American bequest." It's what you have right now. In your bank.

Under your mattress. In your home (if you own it) or in other investments. In unused credit. Small or large (in your eyes or others'), this is your ten bucks.

Now, what will happen if you put some of that into job training or education? What if you invest some of that in an enterprise? What if you vote to send more of it to the local, state, or federal government? How much trust do you have that what you get back will be fair compensation for what you contribute?

This is more than a game, obviously, and the question about how you choose to spend your ten bucks, right now, is a serious one. If you spend another $375,000 to become a physician (that's low-balling what it costs to become an MD), will you be able to come back and practice family medicine in your home town? If you sink $10,000 into a promising IPO, do you feel assured that due diligence has been done on the company and that the offering will be fairly monitored? If you borrow and then spend $23,000 on kitchen improvements, do you have confidence that you'll make it back when you decide to put your home up for sale? When you vote for the governor who wants to increase your taxes to improve the roads, do you trust that the money will be used for road improvement and not for legislative pork-barrel projects?

These are all the questions that face us, right now, and they are all part of the Trust Game. For every one of the decisions that we make about how to spend our ten bucks, there is another participant (or a number of participants) whom we can trust...or not. Of course, neither you nor the other party has complete control over circumstances. You could both lose it all. But in the Trust Game, what you need to know—and what to *find out* if you don't know— is whether you can trust the other participants.

And here's where we get to that other version of the Trust Game that I described, the one where the Wharton professors tried to figure out what would happen if there was outright deception. How do you respond when the other participant takes your money, promises to work with you, then *doesn't?* What happens to your trust level when that other, unseen person lies to you—and as a result, you suffer a direct loss?

The analogies? An educational institution encourages you to incur tens of thousands of dollars in debt without fully informing you of the terms of the loan. The IPO offering that is deliberately inflated to allow preferred investors to get in and get out quickly while others take the hit. The mortgage company that ardently promotes a second mortgage while concealing exorbitant closing costs and variable-rate clauses. The governor who postpones the promised road-improvement bill and instead allocates a windfall of new taxes to a pension-fund shortfall.

Those are more than goofs. Those are deceptions. And, as the Wharton experiment showed, it takes much, *much* longer for trust to be restored when there has been outright deception.

So what *will* it take to restore trust? Not surprisingly, the answer is, "It all depends." If the counterparty in your last transaction behaved—as far as you can tell—with good intentions, and circumstances went against you, that's one thing. If that counterparty, in your view, practiced outright deception, that's another. In the first circumstance, restoring trust is a surprisingly smooth process. But where there's been outright deception, the restoration of trust becomes a long, slow process. And, in fact, may never happen.

Chapter 6

The High Price of Corruption
Free Speech

The designers of the Constitution had a great idea. One of the things they didn't like about the old relationship with King George and England was the "what-we-have-here-is-a- failure-to-communicate" problem. If the colonies had an issue—poor management, lousy taxation policies, unfair restrictions—there was no way to bring grievances to the King, get a hearing, and have your wrongs righted. It is, indeed, a helpless and hopeless feeling when you've got a grievance and can't air it. In the Declaration of Independence, the Founding Fathers complained they had repeatedly, "Petitioned for Redress in the most humble terms," without getting any satisfaction from the King or houses of Parliament. So the Fathers slipped into the First Amendment of the Constitution the telling phrase ensuring the "right of the people... to petition the Government for a redress of grievances."

Not only was this a great idea, it was a morale booster. It suggested that anyone, no matter how lowly in stature or frustrated by the system, could write to their President or Congressman and complain their rights had been violated or that some subparagraph of the tax code was being unfair to them. Lumped in with all the other guarantees of free speech in that amendment, it suggested a recourse that could, if implemented correctly, become a release-valve for a lot of pent-up frustration. If you've got a problem, you have the right to go see someone about it. The Complaint Department might not be open, you might not get the hearing that you want, but this amendment gives you the right to air your grievance and see whether you can get the situation corrected.

But sometimes there are unforeseen consequences. What the framers of the Constitution did not envision was a system where businesses, companies, organizations, and special-interest groups would hire phalanxes of lawyers and lobbyists to occupy plush offices all up and down K Street in the nation's capital. They could not have imagined the highly paid armies of persuasive individuals with ever-expanding expense accounts who stalk the halls of Congress and fill the restaurants, resorts, and luxury hotels with expressions of their largesse. In all likelihood, both Jefferson and Hamilton would be agog at the offsite methods of persuasion being used to pursue the agendas of those who have the money to pay the persuaders. And it's quite possible they would have fretted about the way their carefully worded phrase about seeking redress of grievances had been turned into an all-out, no-holds-barred action plan for lobbyists, consultants, and strategists of every ilk to threaten, cajole, bully, and needle the elected representatives of the people of the United States.

Is Lobbying Corruption?

But wait a minute. Lobbying is perfectly legal. And, these days, lobbyists have to register. And aren't there restrictions on what they can and cannot do? (Guys like Abramoff crossed the boundaries. But guys like Abramoff get caught. There, that proves there are limits! The system works!)

And, in one sense, it does. Anyone can hire a lobbyist. Any lobbyist can register. Any cause can be advocated. Any point of view can be presented. Any grievance can be heard. With lobbying, the citizenry is given access that might otherwise be denied if these voices were silenced. All good and persuasive arguments in favor of the system as it stands.

But in the general public these days, a certain negative attitude has developed in regard to what are customarily branded "the powerful special-interest groups" that appear to control legislation and regulation. There is a pervasive sense that the right to air grievances—so carefully written into the Constitution—has somehow been subverted by the very system that has grown up around it. Today, even the most bullheaded congresspersons will admit their chances of reelection are seriously threatened if they oppose the most powerful special-interest lobbies. While theoretically each of us, as individuals, has the opportunity to petition the government about our grievances, I think most of us are convinced that big interests with big money go first, and that the big-money issues lead the queue of interests that demand to be addressed. Under these circumstances, public morale tends to plummet. It's a bit like that corrupted classroom I described earlier where the student who gets the A's is the one who buys the teacher the most chocolates.

Is this just a perception—or is it real? Is lobbying and political influencing, as practiced today, actually no big deal—just an inevitable consequence of interested parties vying for attention in a prosperous democracy? Or has it, in fact, evolved into a form of corruption?

A Taste of Corruption

Imagine this. You and your caravan of heavily laden donkeys have come to the border of Powderkegstan, a country you have never visited before—but a country in which (you happen to know), the inhabitants are suffering from a shortage of earbuds. It just so happens that the saddlebags of your donkeys are laden with earbuds, so you're quite sure that on the other side of the border there will be a ready market for your product.

At the border crossing, your way is blocked by a young man of haughty mien with a big gun. You show him all your ear-bud-licensing papers, plus a letter from your consulate and his consulate, plus all the immunization certificates for all of your donkeys, but for some reason this gunman looks at you with disdain and declares that he sees many irregularities and is unlikely to allow you to cross this border in the foreseeable future. At that point, you hand him an envelope stuffed with enough cash for the guard to buy another fifty guns, should he so choose. At that point the man's mien becomes less haughty, he sees there is, in fact, a way to correct the errors in your paperwork, and within minutes you are across the border with your heavily laden donkeys and their burden of earbuds.

To anyone who has done business in large portions of the world ranging from the shakiest of republics to entrenched bureaucracies, this has been a well-understood way to expedite business. (Bribery has sometimes been called "the world's third-oldest profession.") If you're dealing with a guard at a border crossing, it may in fact be the most effective way to negotiate with the government he represents. After all, the gun and the uniform give him a power that you don't have, and if you are on a peaceful sales mission and expect to make a hefty profit on the other side of the border, there's no reason why you shouldn't add this small bribe to your expense account and write it off as a cost of doing business.

But if you've got considerably more cash and clout, more earbuds to sell, and you're dealing with a nation having a somewhat more advanced bureaucracy, there is an alternative that may, in the end, be far more effective. When you want to send in earbuds by the jet-load, rather than the donkey-load, it probably makes a lot more sense to lobby the premier or president or parliament of Powderkegstan and arrange for passage of a regulation that will now and forevermore allow for the importation of your company's earbuds. So—what's the better way? The bribe or the lobby? In which country is one more likely to be used than the other? And what's the overall impact on the business dealings within and around that nation?

These were the questions asked by Bard Harstad at the Kellogg School of Management, Northwestern University, and Jakob Svensson of Stockholm University in their paper on "Bribes, Lobbying and Development" in *American Political Science Review*. "The common perception," note the authors, "is that firms in developing countries are more likely to pay bribes to get around regulatory constraints, whereas firms in developed countries are more prone to lobby the government to change the rules. There is also evidence, both across and within countries, that the extent of lobbying increases with income and that lobbying and corruption are substitutes."

Strong words from these professors, considering that most of us tend to think of bribes as one of the lowest of corruptive practices, whereas lobbying—which has evolved from that Constitutionally protected right to seek "redress of grievances"—is well within the bounds of acceptability. So why do Harstad and Svensson utter "bribery" and "lobbying" in the same breath?

It's far more convenient to think of corruption as a criminal act (the Mafia heavies who extort from local business) or as a cost of doing business in a countries like Powderkegstan (where it's just nasty habit). But what if a gray-area form of corruption is closer to home? What if the payoff—not to guys with guns but to guys and gals in suits—now has to be figured into the price of doing business in the U.S.?

Influence Peddling Made Easy

Daniel Kaufman, the former director of Governance at the World Bank and a senior fellow at the Brookings Institute, who studied the kind of corruption endemic to developing countries, noted that it takes a different form when nations become more economically advanced. Writing in Forbes.com in early 2009, Kaufman pointed out that, "One neglected dimension of political corruption is 'state capture,' or just 'capture.' In this scenario, powerful companies (or individuals) bend the regulatory, policy and legal institutions of the nation for their private benefit. This is typically done through high-level bribery, lobbying or influence peddling."

For many years economists who compared relative levels of corruption in various nations found that the U.S. ranked as one of the "least corrupt" nations on the list. But as Kaufman reviewed the statistics, he broadened the definition to include "legally corrupt" manifestations such as "the extent of undue influence through political influence and powerful firms" that could bend government policies to suit their purposes. Adding this measurement, the U.S. was no longer among the top twenty virtuous nations; it was ranked 53rd among the 104 countries surveyed. For examples of how things work in a society where influence-peddling exacts a cost, Kaufman pointed to many of the gaffes that led up to the financial crisis, such as:

• The millions of lobbying dollars spent by Freddie Mac and Fannie Mae influencing key members of Congress to loosen up capital reserve requirements.
• The way the largest investment banks influenced the SEC to relax regulatory measures, allowing them to take on huge amounts of debt.
• The maneuvers of huge mortgage lenders to loosen up regulations by placing themselves under the "lax oversight of the Office of Thrift Supervision."

Made obvious by the hugely enlarged bank accounts held by executives of these companies and the dire consequences for the majority of the American public, influence leveraging resulted in massive damage. We are still trying to clean up the mess. But in the meantime, the fast-and-furious forms of lobbying employed to influence government policies, oversights, and directives have, if anything, only intensified.

In 1990 contributions from the finance, insurance, and real estate sectors to all federal candidates, party committees, and leadership PACs (political action committees) was a meager $61.9 million. By 2008, the year of the crash, these sectors were doling out $497.7 million—nearly half a billion dollars annually—to influence public policy, discourage federal regulation, and keep the doors open for the growth of these lucrative industries. What's even more curious than the 800 percent growth in expenditures is the relatively balanced way in which these funds were spent. In 1990, $32.2 million went to Democrats and $29.7 went to Republicans. In 2008, $251.9 million went to Democrats and $245.8 million went to Republicans. In other words, through a period of the most tumultuous changes in the finance, insurance, and real estate sectors, the institutions that invested in public policy carefully hedged their bets, keeping an eerily even balance between Republican and Democratic parties and candidates.

But it would be distorting the picture to focus on the influence peddling of these three sectors without looking at the contributions of other groups. In 2007-2008 among the top ten donors to national and state politics were organizations like the National Education Association, the Pechanga Band of Mission Indians, Penn National Gaming, Morongo Band of Mission Indians, and Lakes Entertainment. Clearly, many entities have investments in politics, and they represent widely divergent interests—a good indication of democracy at work, albeit with (in many cases) a great deal of money behind this particular form of free speech.

In K Street activities, too, there's ample evidence that a wide range of views are being represented by lobbying firms that are hired to do just that. For the years 1998 to 2012, the top spenders among lobbying clients were the U.S. Chamber of Commerce, the American Medical Association, General Electric, Pharmaceutical Research and Manufacturers Association, the American Hospital Association, and AARP. (Expenditures, during those fourteen years, ranged from $860 million for the U.S. Chamber of Commerce to $270 million for General Electric and $215 million for AARP.) While these are obviously heavy hitters, there is always room for other players in the lobbying game. In 2011, organizations pursuing specific ideological or single-issue goals also hired lobbyists to represent a wide range of causes, from human rights and the environment to abortion (pro and anti), gun rights and gun control, and foreign and defense policy. But in that year, this entire sector spent $140 million when lobbyists for health providers spent well over half a billion dollars.

Obviously there is an enormous disproportion in scale of lobbying investment from different sectors of the economy. But the fact remains, any company, organization, fund, or public cause that wants to participate in politicking or lobbying can do so as long as they're willing to bear the cost.

That said, why has lobbying been so reviled? Why is it that so many voters, both left and right, blame Congressional toadyism on the overbearing influence of "special interests?"

Seeing Is Believing

For most of us, I would guess, there are really two causes for outrage. *Actual* corruption—in business, politics, or government—is something that we generally find distasteful. From the donkey driver who bribes the guard at the Powderkegstan border to the Wal-Mart that meets the *private-needs requirements* of foreign-government officials, there's something rancorous about the whole business of greasing palms. As much as we enjoy Mafia movies as entertainment, we are sickened to discover how individual businesses or industries are subject to extortion by thick-necked men who intimidate and kill. We are indignant when we see that Johnny can get A-pluses by giving the teacher a box of favorite chocolates or that a Representative can pass a land-use deal if given a few acres for his personal wealth-improvement program.

Why the resentment, indignance, or outrage? Clearly, it's the "fairness" thing again—some fundamental psycho-emotional response that gets triggered when we see injustice.

But similar reactions are triggered in the public consciousness when, collectively, there is the *perception* of corruption. When a majority of congressional officeholders subsequently "retire" to highly paid positions as lobbyists or political consultants, it is very hard to ignore the special connection between Congress and K Street. When individuals, corporations, and special interests donate hundreds of millions of dollars to political campaigns, it becomes increasingly difficult to believe those with the deepest pockets do *not* have the greatest influence on policies promulgated by those politicians. When key individuals in all industries move briskly between regulatory agencies and the businesses they are supposed to regulate, *perceptions* of corruption run rampant.

And the American public's perception of government, in the wake of the near collapse of the financial system in 2008-2009, was that the series of government actions were corrupt. The laws that had been passed in the preceding years so favored business, insurance, and real estate (the sectors making the biggest lobbying contributions) that they were obviously getting their way. Agencies designed to protect the public interest were so overburdened, underfunded, and mismanaged that their regulatory powers were negligible. During the panicked attempts to rescue banks and financial institutions from massive default, those who were "made whole" again by the bailouts were the world's largest financial institutions, and their executives by extension, not the homeowners whose lifetime savings were eviscerated. Given the outcome of closed-door decision-making by Washington insiders meeting with leading representatives of Wall Street's largest firms, the nation's economic interests were salvaged, and, yes, those who ran them remained wealthy. This intended or unintended consequence left a very bad taste in the mouths of many.

The fact is, perception of corruption diminishes us all. We don't appreciate having a Congress where it appears that influence is for sale to the highest bidder. We don't like it when protectors of public safety—whether cops or meat inspectors or atomic energy commissioners—seem to look the other way. And we really don't like it when the dollars meant to build roads, schools, and hospitals go to fund benefits packages. It's stuff like this that makes us feel as if we're living in a third-world country or (perhaps worse) some spinoff of a centralized Soviet bureaucracy—those undesirable places where corruption is rampant. When we catch the stench, corruption makes us wrinkle our noses.

The Reality of Giving

How ironic, then, that so much of this perceived corruption goes back to that idealistic phrase—"the right of the people...to petition for redress of grievances." Fortunately, that right has not gone away. You can still email your Congressperson. You can create The Committee for the Promotion of Kind Acts and raise money for that group, hiring a K-street firm to lobby for Kind Act legislation. You can buy TV ads in support of Kind Act candidates, and if Mean Act legislation is passed, despite all your efforts, you can lobby for its repeal. So, theoretically, you have many avenues of recourse as you seek redress for your grievances.

But in reality something quite different is going on, and perhaps it is this that makes us most uneasy with the current state of grievance-addressing in American politics. Though none of us can say exactly what the framers intended in the original wording of the Constitution, we do know how it would have to be re-worded, today, to represent what's really going on:

"Those who are funding political campaigns and lobbyists shall have unlimited options to promote candidates, support and maintain them while they are performing legislative acts in the funders' self-interest, to signal approval or disapproval of said candidates through the sustaining or withdrawing of financial support, and to threaten such candidates with loss of office should they vote against favored legislation or government action."

This is not something we would be proud to see as part of our Constitution, but surely it's a forthright statement about the current state of affairs. It more accurately reflects the way things actually work, today, among our nation's politicians and the bevy of corporations, non-profit organizations, and various special interest groups that are working hard to promote their respective agendas.

But why decry the shortcomings of this highly evolved system?

To answer that question, we have to get into the heart of that entrepreneurial salesman with his donkey-loads of earbuds on the border of Powderkegstan. All his hopes for the future lie on the other side of that border. Over there, just beyond the guard with the gun, are potential customers just waiting to buy earbuds. The donkey driver may be carrying his lifelong investment in his saddlebags, but it will all come to nothing if he can't get to the other side of the border. And the only thing that stands between him and this promised land of eager earbud buyers is a greedy border guard with his outstretched hand.

Herein lies the greatest problem with corruption. It makes us, with whatever hopes we carry, subservient to those who can stop us at the border of opportunity. Whether you're the corner grocery being forced to make a weekly donation to the Goodfellas Fund or GE making an annual contribution to a friendly candidate, the act of contribution means you're ceding power to a higher authority—someone who can do something for you. You're not petitioning for redress of grievances. You're buying access, or influence, or favorable action.

And the question is, do we want to be perpetually put in that kind of position? Can we tolerate this kind of corruption?

Of course we can and we do. But it's not a good strategy. Inevitably, it can only lead one direction—to further erosion of trust incurring, as a consequence, the higher social and financial costs that must be borne as trust erodes. The political contributions and the lobbying—like the bribe to the border guard—may indeed become written-off as the "cost of doing business." But they are lousy investments in short-term results. And in the end, they only generate greater distrust.

Chapter 7

Side Effects of Hope
The Devastation of Despair

We get to choose between hope and hopelessness, even in the most abominable situations. It was psychiatrist Viktor Frankl, survivor of four Nazi death camps (including Auschwitz), who wrote, "Everything can be taken from a man but one thing: the last of the human freedoms—to choose one's attitude in any given set of circumstances to choose one's own way."

I hope none of us will have to make those kinds of choices under such horrific conditions—and yet, every day, we respond to the physical, political, social, and cultural challenges that we face. And no two people are likely to react in the same way.

When Charles Hopper lost his job at Lehman Brothers in 2007, he was earning a seven-figure income with salary and bonuses. He and his wife had a 3,200-square-foot, three-bedroom house in Cos Cob, Connecticut, a suburb of Greenwich, with an estimated worth between $1.2 and $1.5 million. After Hopper left Lehman Brothers, he worked a short time for Citigroup but was laid off in 2008 after the financial crash. In August 2010 he landed a new job at a firm that customizes hedge-fund portfolios, where he earned $150,000 a year. On May 14, 2012, overwhelmed by financial problems, Hopper went into the garage of his home and hanged himself.

When Brianna Karp was laid off from her job at Kelley Bluebook in 2008, she was making $50,000 a year and living in a little cottage in Orange County, California, that she rented for $1,500 a month. The recession hit southern California with full force, there were no jobs, and Brianna's unemployment checks came to $1,800 a month. She realized she had to move out. During the months that followed, she lived in a trailer in a corner of a Walmart parking lot. Her days were spent at a nearby Starbucks where the baristas got used to the sight of the homeless young woman hunched over her laptop as she applied for hundreds of jobs. Finally she was offered a job as an executive assistant at a five-person web design company, making far less than she had at Kelley but more than her unemployment checks. She accepted, meanwhile beginning a lively online correspondence with other homeless people dealing with similar issues, launched a website called girlsguidetohomelessness.com, and wrote a book about her experiences as a homeless person surviving one of America's deepest recessions.

Neither of those portraits is complete; neither one tells the whole story. But here is one of the mysteries about that ephemeral quality we call hope. Both were individuals caught in a downward economic spiral that neither of them could control. Both lost their jobs—with all the financial hardship that implied. Both experienced a drastic change in expectations. One lost hope. The other did not.

Why?

If we knew the answer to that question, I'm sure our attempts at social engineering would work a lot better. Given exactly the same social conditions, there are some who will maintain their faith in the future while others will give up. During times of economic instability and upheaval, some people with virtually no assets manage to maintain optimism while, at the other end of the spectrum, are individuals with great wealth who fall into despair. None of these responses can be totally predicated on relative affluence or poverty.

And yet...as we all know, an economic bust creates huge side effects. America saw it in the thirties, and we are seeing it again today. The ripple effect of an economic crash goes out in all directions, disrupting the lives, wellbeing, confidence, trust, and even the health of those caught in new uncertainties. Those who drop out of the work force must fight ever-tougher battles to maintain dignity and preserve their self-image as worthwhile individuals having real value to society. We know what it costs in terms of self-image when a proud family provider begins using food stamps or visits the neighborhood food bank for the first time. We understand what it means to someone like Charles Hopper to be stripped of the power and prestige of a high-paying job, or to someone like Brianna Karp to find herself "branded" as a homeless person. These are circumstances in which we and our neighbors find ourselves having to reach farther, and hold on more tightly to every straw of hope that we can grasp. The fact is, despite all our best efforts, hopelessness can take on epidemic proportions. It is just one of the side effects of what we so accurately label as hard times.

Hope in Hard Times

One of the pervasive illusions, throughout history, fiction, and mythology, is the notion of a promised land. We know what it was for the Jews who followed Moses– a locale just on the other side of the Red Sea where there were no pharaohs practicing slavery and nothing to stand in the way of their exercise of a monotheistic religion. We know what "promised land" meant to Columbus and those who immediately followed in the wake of his tiny ships—a continent that would produce gold, spices, and other sources of great wealth. To the settlers of America, from lowly Pilgrims to mercantile-minded land grabbers given rights by their kings, the promised land was fresh soil ready for the tilling. Westward exploration opened up the rich farmland stretching all the way to the Mississippi, where forests, fields, lakes, and streams gave agrarian-minded settlers the promise of endless bounties of food. Promises, promises.

Of course, what each of those people found when they got there was quite different from the gilded images planted in their imaginations by charismatic storytellers, bragging travelers, quixotic dreamers, romantic-minded religious leaders, or highly imaginative power brokers. Moses certainly had great leadership abilities, but as a travel planner he was a washout. Whether or not the seas actually parted, we do know that the land where the Hebrew people arrived was not the kind of place you'd want to send your son for summer camp—dry, hot winds, scarce vegetation, and vast drought-stricken expanses. And in those few Mideast locations where conditions favored habitation, there already lived groups of hostile-minded residents who had no intention of giving up their oases without a fight. Moses' promises—however he framed them—were about as substantial as a Florida land deal.

Columbus fared no better. Raising capital and crew for his ships was child's-play compared to the adversity he faced upon arrival on the beaches of tepid, mosquito-ridden islands inhabited by small groups of natives who in no way resembled the rich rajahs who were supposed to be sitting around in palaces layered with gold and encrusted with precious gems. Desperate for some return on investment, Columbus tried to make the best of it—substituting novelty items (slaves, herbs, stories) for actual riches, but his three long, wandering voyages involved more shipwrecks than shipments and more mutiny than remuneration.

And so it goes. We may love the tales of the Bradfords in Massachusetts or the Ashley-Coopers of South Carolina, but their respective promised lands were more hellholes than Valhallas. Conditions turned deadly, disease was persistent, winters proved nasty, food ran out, Indians threatened. Not to mention the slew of petty jealousies, rivalries, and disputatious clashes of creeds that occurred among people who arrived on these shores looking for the realization of their dreams and found, instead, a new and more creative form of struggle for existence.

As for the westward expansion, if you have issues with damp, cold, storms, bugs, spiders, and you don't go for an all-winter diet of dried corn and tuberous vegetables, you would definitely not be motivated to pack all your belongings on horses and wagons and head in a general westward direction. Yet people did. And while their experiences have few of the heroics depicted on celluloid, they did have their illusions of a land of plenty, just on the other side of tomorrow and tomorrow and tomorrow. And the ones who stuck it out did, in time, see some results. (The ones who didn't, of course, sickened or starved or froze or forsook the hinterland for familiar grounds.)

In more recent days, on a continent explored and settled and divvied up among various private owners and government entities, the promised land does not lie *out there* but much closer to *right here*. Of course, now that we're older (as a society) and wiser (presumably), we have a semi-ironic designation for that promised land. We call it the American Dream, hoping, that by doing so, we acknowledge that the land deal we promised ourselves is not, in fact, a promise, but a dream. With house, family, car, a nearby quality grocery story, an upscale mall within easy driving distance, we expect to have it all.

But this American Dream, like all the promised lands that came before and the ones that will come after, carries its own baggage of expectations. If Moses had been a little more forthcoming about the land deal he was selling his people, he would have noted that you only get the promised land if you do a lot of hiking over dry land, survive some unmentionable conditions, and are prepared to fight for what you get. A more honest Columbus would have come back from Hispaniola with a lot of dire warnings rather than a sales pitch. And any of the travelers setting out for American shores really should have been provided with a complete brochure explaining that conditions were rotten, winters hard, luxuries few, and there were no guarantees of safety for one's health, family, or personal belongings.

And so it is with the American Dream. Fine, believe it if you want. Go for it. But if you're going to stick it out, you have to be prepared to deal with the unwritten terms of this contract. Those terms being: anything can happen, and you can lose it all. There are no guarantees.

So—given those terms, are you still willing to hope for your promised land?

Infectious Hopefulness

The answer, for most of us is, sure. Yes. Why not?

And once we've decided (instinctively or consciously or desperately) to *hope*, our perceptions become skewed and our reasoning turns unreasonable.

In a talk given by Dan Gilbert at the TED Conference in 2005, the Harvard psychologist teased the audience—a group of intellectuals and innovative thinkers gathered to discuss the state of the world—with an arcane formula for rational human action that was first proposed in 1738 by an early interdisciplinary Dutch thinker named Daniel Bernoulli. Bernoulli's formula, as Gilbert points out, was actually quite simple: "The expected value of any of our actions— that is, the goodness that we can count on getting—is the product of two simple things: the odds that this action will allow us to gain something, and the value of that gain to us."

While Bernoulli's formula may have validity in theory, Gilbert spends the balance of his talk proving to his audience, quite convincingly, that nearly every decision we make is *not* based on anything so rational as the Dutchman's precise logic would suggest. What we act upon, instead, is perception. And perception is a very powerful manipulator of decision making.

To make his point most emphatically, Gilbert uses a lottery as an example. But instead of looking at the lottery as a gaming field where everyone has a fair and equal chance to win (as I described in Chapter 1), Gilbert looks at the enormous *improbability* of any one person winning, and then asks, "Why in the world would we ever play the lottery?"

The reason, of course, is there is *a winner*. When the winner is announced, it's *stop the presses!* All media focus on the individual, or group of individuals, who are weeping with joy. But, says Gilbert— and this is important—a typical lottery might have one winner for every one hundred million losers. He calculates that if *everyone* who participated in the lottery was interviewed on TV for thirty seconds, you would have to watch TV for nine and a half years, nonstop, while the one hundred million losers were interviewed. Then, for thirty seconds, you'd get to see the winner weep for joy.

If that happened every time there was a lottery drawing, how many people, do you suppose, would continue to play the lottery?

We look at the bright side. We look at the possibilities. We take crazy risks (crossing the Red Sea, the Atlantic, the American plains) because we have this perception that somewhere, at the other end, when we arrive, we'll be weeping for joy.

And this is why Bernoulli's formula is such a tease. When you look at the behavioral kookiness in human history, it appears that we are consistently incapable of applying the most basic qualitative and quantitative evaluation to our actions. We just can't properly measure "the odds that this action will allow us to gain something, and the value of that gain to us." Call it hope, wishful thinking, faith, delusions of the promised land—whatever it is, the "go-for-it!" in our brains makes us a bit crazy.

Where Go-for-it Leads to De-fi-cit

I encountered this, first hand, when I was a brash youth of thirty years, just getting acquainted with a board of directors in an organization that was specifically and solely designed to carry out good works. The organization was a charity founded to help distressed travelers in Washington, D.C., and its primary purpose was to provide assistance for travelers and immigrants—especially those with little money and limited language ability—who were arriving daily in the capital region. It was a privately funded organization with over a million dollars in its endowment. Soon after I joined the board, I learned that the Society was spending down its endowment faster than the funds were being replenished by gifts and donations.

What to do? The acting director proposed a novel solution. What was required, he said, was spending on an even larger scale. He said we needed some very high-profile, costly projects that would draw attention to the work of the Society. Theoretically, new donors would be inspired by these public acts of good work and would knock down our doors with additional contributions.

The director was inspiring, and his presentation created a kind of euphoria in the room that was hard to dispel. I watched as several other members of the board—all with more seniority and experience than I—one by one gave their assent to new plan. Sure, we would be outspending our endowment—we were sure to end up in the hole—but our business was charity, and if we did not do the work our society was designed to do, how could we hold our heads up?

Unfortunately, thirty-year old Farr didn't have the sense to keep his mouth shut. "Let us suppose," said Farr, "that Mrs. Farr and I had the good fortune to make $100,000 this year and after looking over our budget, we decided that we would really like to give away $120,000 to the needy in our community. And suppose we went ahead and did that and were bankrupted. Could I have a show of hands? Is there anyone here who would *not* judge Mr. and Mrs. Farr to be horses' asses? I'm sorry, this is not my way of doing things, and if this is the board's decision, I will have to resign."

At that point, damage done, I prepared to leave, but a seventy-three-year-old member asked for a twenty-minute recess to meet with the chairman of the board. So they did. The discussion was brief. At the end of it, the chairman offered to step down, and for better or worse I was elected chairman. The directors reassembled, and my nomination was unanimously approved. I was speechless. As usual, my impulsive mouth had gotten me in trouble (I've learned better control), but I also had a mandate. I fired the director and hired a new one. Some people were laid off—with much protest from members of the board. (*We were a charitable organization! How could we do that to people?*) It was not kind, not gentle, not sensitive, not in the spirit of philanthropy. But within six months the charity was in the black again, and we could continue our work.

Would I have liked to provide aid to every single traveler-in-need who arrived in the capital? Yes. Would it be charitable and generous to do so? Yes. Was it possible? No. And if we attempted to do that, depleted our funds, and ran into the red, there was no way we could keep the charity going and continue to fulfill our purpose.

Needless to say, I have been reminded of this episode numerous times in recent years. As the concept of deficit spending has gone into high gear—in government, business, and our personal lives—it's not hard to see what keeps propelling us forward from the black into the red. Out there on the horizon, just beyond reach of all of us, there is the promised land. It is a land where there is no poverty. Where every child is fed and educated and granted the means to live in a reasonable degree of comfort. In the promised land, business risk is buffeted by guarantees that, even if your business fails, there will be a soft landing or a kindly bailout. In our personal lives, the promised land is that American dream we refer to so often—yeah, the one with the home and kids and car and (now) with an iPad and cell phone in every brand-name handbag and briefcase. As we all know (just regard the current deficits), our hope has been replaced by promises, and in the process of fulfilling promises, we totally neglected Dutchman Bertoulli's judicious formula. We know the value of these things we all hope for. But what are the odds of getting those things if you can't pay for them? Not very good. The fact is, if we think we can keep taking in $100,000 a year and spending $120,000 a year—just because it's the charitable thing to do –we're probably horses' asses. (And I do wish there were a more polite way to put it.) (But there isn't.)

Rights vs. Hopes

Something else happens in our brains when we have the illusion that we have not just the hope, but the right, to win the lottery. With heightened expectations and the promise of good things to come, we all become a little more susceptible to flimflam.

Here's how that works, as recent history has shown. Say you have been in the workforce a couple of dozen years, and you're pretty loaded down with expenses and obligations, but then along comes someone who offers you a house for no money down. Now, you weren't born yesterday, and you know a thing or two about the world, and usually your crap detector is working pretty well, but in this situation your keen senses seem to go into a state of hypnotic rapture. Why? Because that dream—owning your own home—could be, just could be, within reach. So...let's hear what the guy has to say...

And, right there, you become susceptible to all the powers of persuasion that flimflam artists know so well. First of all, says this purveyor of deals, you deserve the house because you're an American and you've worked very hard a couple of dozen years. Secondly, you can afford the mortgage payments because we will start you at a very low interest rate which will, indeed, go up after a while, but so will your salary and—well, maybe your spouse's salary, too. And, besides, by the time your mortgage payment goes up, your house could be worth two or three times what it is today. So...what's holding you back?

Now it's true many of the buyers of low-down-payment or nothing-down homes have been portrayed as victims, as fools, as greedy, or as dupes. Without seeing the individual motivations in individual circumstances, we don't really know—and, in any case, we don't get very far attaching labels to people. What we do know, however, is that every one of them signed a document, an agreement to pay in the future for the property they would live in or invest in. And at the moment of doing that, whether they experienced a feeling of dread (*Can I really do this?*) or a feeling of glee (*I'm going to make out like a bandit!*), we know that each of these individuals was susceptible to the delusion that the promised land was within reach.

Nor was this kind of delusional thinking confined to first-time homebuyers. Different folks have different promised lands. For the rising star in AIG dealing in the newfangled insurance instruments known as Credit Default Swaps—the guy or gal who already had three homes and five cars—the promised land was a year of record-breaking profits that would deliver a bonus like none ever seen before. For this opportunity, he could make wildly, insanely optimistic promises that his company would never need to ante-up for the tranches of mortgages that he promised to insure. Deluded by visions of his promised land, he could make himself believe that all the credits and mortgages he insured were sound, that the real estate market could keep going up indefinitely, and that his company could somehow find the money somewhere to prevent a total collapse of the intertwined obligations. These individuals, too, have been called by many names, but whatever their character flaws or innate virtues, all were susceptible to a dream of great wealth (their promised land). And they effectively scrawled their signatures on deals they could not possibly keep.

Yes, it's a lot more fun living with hope than living without it. But one of the side-effects of hope is that it's the very thing that can put us in the danger zone.

Dealing with Side Effects

So this is the question Dan Gilbert asks: How can we make better decisions for ourselves and our society? Unlike a lot of other animals in the food chain, we really should be able to apply Bernoulli's law—estimating the odds that we might succeed and estimating the value of our own success. In fact, Gilbert ended his TED speech with a very big exclamation point about the need to do this: "If we're not here in ten thousand years, it's going to be because we could not take advantage of the gift given to us by a young Dutch fellow in 1738, *because we underestimated the odds of our future pains and overestimated the value of our present pleasures.*" (Italics added by author.)

Fortunately the future of human civilization does not rest on one person's decision to buy a house for no money down or another person's decision to insure tranches of worrisome securities. But the mistakes we're making are similar. We want the promised land, and we want it now. And the problem with this attitude is the same problem that has always plagued those who go in search of that land. A, it's a lot harder to get there than we originally thought. And, B, once there, we're going to find that it's not all it's cracked up to be.

The neo-homeowner who bought his dream house did not anticipate the collapse in housing prices or the clobbering of the job market. It was a nasty shock to realize that the mortgage would have to be paid whether the housing market tanked or flourished. The mortgage would be owed whether or not the homeowner was employed. And if there was a paragraph in the mortgage agreement referring to the lender's right to escalate interest, that clause would go into effect irrespective of whether anyone in the buyer's family was getting commensurate raises in income. True, the terms of this agreement may appear draconian: there is no provision for relief should the economy turn against the neo-buyer. But on the other hand, there was nothing forcing the hand of the buyer to sign the contract—nothing, that is, save the powerful magnet of getting what we want *now*.

And the wheeler-dealer at AIG? What was his handshake with destiny? We know he would not have gotten to the position he's in unless he was a remarkably talented individual—talented, that is, in all the ways that made him part of the meritocracy. As a youngster, he competed most successfully against other classmates. In graduating from college and landing this job, he showed all the persistence, dedication, and drive that are required of someone going into a position of responsibility. Over time (long or short) he was increasingly trusted to handle the kinds of sums that would exceed the lifetime income of most American families. And in compensation for his intelligent application of skills, his earning power (for the company), and his loyalty, he began making the kind of salary that could support eight or nine ordinary families and the kind of bonus that could buy several additional houses on Park Place or Boardwalk.

With meretricious performance and heightened responsibilities, however, there also came increased expectations—not only his own but also from those around him. To fulfill these expectations, he would have to be on the lookout for opportunities. When opportunities came within reach, it was imperative that he seize them. And so it happened that when a brand-new financial instrument, like none ever seen before, came down the road (in the form of credit default swaps), he snatched it, ran with it, was honored for his initiative and diligence, increased his salary, raked in his bonus, and, to all appearances, arrived at the promised land.

This, then, is another side effect of hope. It raises expectations. It shimmers with promise. It gets us to try things we would not ordinarily try; to reach for things beyond our grasp; to make deals we would not ordinarily make. And after doing all that, some will be resoundingly punished for ignoring Bernoulli's equation, acting on hope that far exceed the chances of getting what we want. But, all too unjustly, others will be richly rewarded for taking the risk.

It's a fickle process. We know that. And yet we scream for a measure of fairness. We shudder at the sight of abandoned houses and shake our fists at the financial risk-takers who threatened us with economic Armageddon and then got bailed out. Like those lottery players I mentioned at the beginning of this book, we're willing to cope with the knowledge that our chances of winning are very slim, but we hate it when we know the game is rigged in someone else's favor.

And, right now, a lot of us who are looking for the promised land resent the way the game is rigged. Because, the fact is, some of us have to bear the consequences of our actions. And others don't.

How did this state of affairs come about?

Chapter 8

Responsibility for Consequences
Escaping Blame

1. If you drink five martinis and then drive home along a winding road at 78.6 miles per hour, are you responsible for the consequences?

2. If you go into a bank to make a withdrawal from your account and decide, instead, to hold up the teller at gunpoint, will you have to deal with the results of that choice?

3. If you collaborate with the nation's largest banks in fixing the interest rates paid to municipalities all around America, will you end up in the hoosegow or on Easy Street?

The answers to questions 1 and 2 are what we aptly call "no-brainers." Choose to drive drunk, and of course there are likely to be consequences. Rob a bank in broad daylight, and there's probably jail-time in your future. But number 3? Well, that's a bit tougher to answer. Because recent history shows there are not likely to be any consequences for you, personally, if you are involved in a rate-fixing scheme with a bunch of other banks to ensure noncompetition. Not only that, there are quite a few other things that you can do without too much fear of consequences if you're a Wall Street honcho in the right place at the right time.

I'll give a spectacular example of that in a moment.

But first, let's consider what we mean by "consequences."

In examples 1 and 2, above, the consequences come in different shapes and sizes.

Take #1. Being a person with some life experience, you probably know pretty well what's likely to happen if you drink five martinis in quick succession and then place yourself behind the wheel of a car. One consequence could be death or maiming of yourself or others. Another could be major or minor destruction of your or another's car or property. And there could be legal consequences if you're caught. If your mind were clear, you would have some ability to measure these probabilities and make a decision in your own best interest.

Case #2, robbing a bank, is one of those situations where the short-term consequences (a lot of cash) may be very tempting, but the long-term consequences (lengthy jail time) make the opportunity seem far less tempting.

Now, the question is, in Case #3, what combination of personal and/or legal retribution is likely to be exacted as a consequence of your decision to get involved in a rate-fixing scheme? If you have great powers of empathy and put them to use, you might be able to imagine the consequences in human terms—that is, the discomfort or desperation of those town officials who placed their money in your hands, paid your fee, and lost millions of dollars because of your decisions. If your wrongdoing happened to be detected by some legal agency entrusted with spotting fraud, there's a small chance you might have to pay a fine or even do a little time. But given the current lack of empathy among you and your peers, and the laxity of law enforcement by watchdog agencies, there's a much greater chance that the consequence, for you, will be Easy Street rather than the hoosegow.

Case in Point

In the U.S., municipalities raise money for special projects by issuing bonds. Once the bonds are purchased, the funds are controlled by officials of the town, city, or county who spend down the balance over the life of the project—putting up a municipal building, creating parks, improving roads and bridges, building schools. But the funds don't remain idle. The idea is to invest these funds for the highest possible rate of return.

Those earnings can be quite significant if the municipality has millions of dollars to invest. So, to make sure municipalities get the highest possible return on their money, they are required to approach at least three different financial institutions and get independent rate-bids on the funds. It's a rule that's well designed to make the most of taxpayers' dollars. If the three rate-bids come in at, let's say, 5.0%, 5.4% and 5.8% from banks A, B, and C, the township will naturally go with the highest bidder (Bank C) to get the greatest return on investment. While the difference of a few tenths of a percent may seem trivial, over the life of a multimillion-dollar loan, it can represent millions of dollars in additional income for the municipality. As for the technicalities of getting separate bids and presenting them to the municipalities, that's all handled by a broker.

But suppose Boggs the Broker—who does work for a lot of municipalities across the country—is in cahoots with Bank A, Bank B, and Bank C to make sure they each get some of the business. If he plays his cards right, and everyone understands what's going on, Boggs can make sure the bids come in at 5.0%, 5.1%, and 5.2%. Now that's a huge saving for Bank C, because instead of paying 5.8% (which they were willing to bid to get the business), they only have to pay 5.2%. They still get the business! They save .6% over the life of the loan! And they are so grateful to Boggs, the Broker, they arrange for his firm to get a nice reward.

Next time around? Well, same scheme, but this time, Boggs gives Bank A the chance to be the high bidder (by a hair). So the business gets nicely distributed among Banks A, B, and C—with each of them getting to pay less interest than they would otherwise pay—and each, in turn, rewarding Boggs the Broker for his good deeds. The losers, of course, are the municipalities who, in aggregate, are being cheated out of billions of dollars in interest earnings because of the collusion between Boggs the Broker and the three big banks.

Nah—that wouldn't happen.

The big banks would never go for it.

Someone at Boggs the Broker would say, "This isn't right!" And blow the whistle.

The banking execs, concerned about having their reputations bludgeoned, would step in and say, "Stop this now. Enough!"

The Feds would catch on right away and put a stop to it.

Well…as it happens, the Feds did catch on, but it took them a long time, and meanwhile the banks made out like, hate to say it, bandits. Hard-pressed townships around the U.S. were cheated out of millions of dollars of interest-earnings revenue, and guys like Boggs the Broker became very rich.

As for consequences?

Finally, there was a trial, and you could have sat in on it if you had almost infinite patience and the desire to listen to lawyers drone on and on about incredibly boring financial data. The trial, *USA v. Carollo,* began on April 16, 2012, in a federal court in Manhattan. As reported by Matt Taibbi in that most liberal of all liberal rags—*Rolling Stone* magazine—the defendants were three gentlemen who worked for GE Capital, the finance arm of General Electric. As Taibbi sat through the long days of the trial, he realized that for this scheme to work, all the big players had to be involved in some level. That meant not only GE Capital but also J.P. Morgan Chase, Bank of America, UBS, and Wells Fargo.

"By conspiring to lower the interest rates that towns earn on these [municipal bond] investments," Taibbi concluded, "the banks systematically stole from schools, hospitals, libraries and nursing homes – from 'virtually every state, district and territory in the United States,' according to one settlement. And they did it so cleverly that the victims never even knew they were being cheated."

Taibbi quotes a typical conversation heard by the jury in which the representative of GE Capital said he was willing to bid "503, 4" on a $219 million bond from Allegheny County, Pennsylvania. In other words, GE Capital was ready to pay an interest rate of 5.04 percent. After "discussion" with the CDR broker, however, the final bid would come in at 5.00 percent. In other words, during all the years that followed, as the account was being drawn down, GE Capital would continue to pay out at a fixed rate of 5 percent instead of 5.04 percent. CDR was also a winner because the firm got a fee for brokering the deal. The only loser was Allegheny County, Pennsylvania, which got significantly less income from its investment as a result of the lower interest rate.

And so it went, year after year for a least a decade, possibly more. The losses to municipalities are now, literally, incalculable. The bidding system meant to protect the assets of U.S. townships consistently favored the banks engaged in bid rigging.

And the consequences? After they agreed to assist in the government's case, UBS, Bank of America, Chase, and Wells Fargo also agreed to pay $673 million in restitution—though dozens of other banks may also have been involved in the scam. And the three defendants in *U.S.A. vs Carollo* were sentenced to three to four years in prison, "for defrauding cities and the U.S. Internal Revenue Service." A small price to pay for billions of dollars in revenue.

Who's Responsible?

Of course, since Taibbi is a crusading reporter working for a liberal rag, it stands to reason that he would portray the trial he witnessed as blatant evidence of run-amok capitalism and compare the shenanigans of shrewd bankers and corrupt price-fixers to the tactics of the Mafia. The young reporter's outrage sizzles in every paragraph. Okay, but behind the prosecution of the case lay six years of work by the FBI—which is not exactly a bastion of liberal activism. The lawyers defending the case were the best that money could buy—not exactly candidates for Occupy Wall Street. And the jurists who finally understood the depth and breadth of the price-fixing, and returned a verdict of guilty, were mostly elderly ordinary citizens scrutinized by prosecutors and defense attorneys in the usual thorough manner.

What were the consequences for the players? A minuscule fine against the banks. Maximum sentences recommended for the bid-fixers that put this in the category of a second DWI offense. No wonder the media, left and right, are yowling about what's rotten in the state of Denmark.

Only it isn't Denmark, it's here.

And when things are rotten on this scale, we can't seem to figure out who's responsible.

There are some clues, however.

One can be found in a short bit of dialog in the transcript of a conversation between Doug Goldberg and his boss, Stewart Wolmark, at the brokerage firm CDR (at the center of the rate fixing scheme). The conversation occurs right after the young broker (Goldberg) has concluded a deal for one of the banks and has arranged for a kickback to CDR.

His boss, Wolmark, says, "Hey, congratulations. You got another one."

Goldberg replies, "Yeah, thank you. Thank you."

"You're hot!" Wolmark compliments him.

"I'm hot? Hot with your help, sir."

Shall we call this a mentoring relationship? If so, it has entirely the wrong flavor. Where is the boss reminding his protégé that he works for a firm that needs to maintain its reputation? What fundamental guidelines about their business have both of these guys forgotten? Why is Wolmark giving Goldberg a big pat on the back for carrying out a deal that, when known, would completely undermine their credibility?

It's one of those situations where there's no parent in the room. The mentor is not a mentor—he's just another kid getting away with something. And his junior partner in these antics is, obviously, a fast and eager learner. Neither one is up to handling the responsibility they've been given. They're just kids with their hands in the cookie jar.

You could say that it's the responsibility of the FBI and the Justice Department to arrange for consequences, and in this case they certainly did. But six years of investigation and evidence-gathering is a long time, especially for a scheme that is being carried out on such a massive scale. Even a simple open-and-shut case like this one (where there were lots of informants) is not so easy for jurors to understand, much less arrive at a verdict.

And what about all those mayors and town managers and county treasurers who got duped into thinking the banks were submitting clean bids for the business? Surely, if they'd been a little more savvy they could have questioned the process, or demanded better service, or detected some malfeasance. Shouldn't they have caught the first sniff of corruption and done something about it? After all, they're the ones responsible for guarding the taxpayers' funds.

Is it entirely up to Wolmark and Goldberg to decide whether they want to drink the five martinis and get behind the wheel of the car? Isn't someone supposed to stop these guys, teach them the rules, slap them with rulers early in the game so they don't get all carried away with the fun they're having when they're "hot?"

The answer, I'm afraid, is "probably not." These guys are so smart, and they're in such an advantageous position, and they get so charged up with the whole thing, that it would be almost a fulltime job by a fulltime police force comprised of even-smarter guys just to put a hold on their antics—much less stop them cold. That's why the great Wall Street disaster of 2008 didn't have the consequences we'd hoped for. It didn't bring about greater transparency. It didn't change the way hedge funds work, or the way banks protect their assets, or the risks that traders take. None of the innovative financial instruments were outlawed, and as for the derivatives market, it only grew larger. And the guys who are doing these deals—many who are much richer, more powerful, and more dangerous than Wolmark and Goldberg—still take no responsibility for the consequences of their actions.

What's Behind It All

Not surprisingly, I'm not the only one who has tried to figure out what lies behind the sorts of ethical lapses—not to mention the stupidity—of the financial activities that have made Wall Street

Bankers among the least-trusted groups in American society. Among others who have puzzled over their rank malfeasance is commentator Christopher Hayes who, in *Twilight of the Elites: America After Meritocracy*, conducted wide-ranging interviews in an attempt to understand what lies at the root of so much comfortable corruption. As the title of his book suggests, Hayes blames the "meritocracy" that has come to dominate politics, industry, finance, and the media in this country. Within the folds of this meritocracy are the self-appointed power barons who, through wealth and privilege, have lost touch with any real feelings of empathy for other members of society. Hayes observes that "three decades of accelerating inequality have produced a deformed social order and a set of elites who cannot help but be dysfunctional and corrupt."

Hayes use of "meritocracy" is as something of an end-stage condition rather than a process of advancement driven by talent, merit and hard-work. Whether or not you agree with his leftward leaning perspectives, he makes an excellent observation regarding the detachment of privilege. Marie Antoinette comes to mind saying, "Let them eat cake!" What happens when wealth, position and privilege insulate those who have achieved (and those who have had achievement thrust upon them) to the point they no longer can relate to those of us still consumed with our life's work? Moreover, Hayes posits that detachment and insulation can lead to insidious actions that undermine the foundations of capitalistic accomplishment and possibility.

Pointing to the upper one percent who have so clearly pulled ahead of everyone else—both in economic terms and in the power they exercise—Hayes proposes that a number of factors contribute to the rise and dominance of a meritocracy. The first, of course, is money—which not only gives individuals and institutions the power to act as creditors but also allows for the buying of influence at all levels of government. Political influence, in turn, leads to the making of laws, regulations, and legislation that favor the wealthy. Hayes cites the 2005 study by Princeton political scientist Larry Bartels who analyzed the favoritism shown by senatorial voting records. "In almost every instance," Bartels concluded, "senators appear to be considerably more responsive to the opinions of affluent constituents than to the opinions of middle-class constituents." As for those at the bottom of the economic ladder, Bartels could not detect they had any influence at all on roll-call votes.

Another source of power among the "elites" comprises networks that were well identified by C. Wright Mills in *The Power Elite* published over fifty years ago. "In political, economic, and military circles," wrote Mills, there is "an intricate set of overlapping small but dominant groups [that] share decisions having at least national consequences." To these circles we would have to add, today, the academic ones—such as professors of economics who move with effortless ease from the university to business to government and back again. "The members of today's elite," Hayes notes, "have never been farther from the median worker and closer, in the literal sense, to their fellow meritocrats." Inevitably, a lot of ethical dodging and weaving is required as, for instance, leaders of industry go on to assume power positions as regulators of those same industries; as former congresspeople move to K street and begin lobbying colleagues; as leaders of Wall Street firms graduate to roles as presidential advisors; and so on. If anything, the power elite is even more incestuous today than it was in the post World-War II years when Mills identified the intertwining networks that created an ad hoc governing body.

Finally, Hayes points to a wide variety of what he calls "1 Percent Pathologies" that seem to be part of the psychological make-up of those within the meritocracy. As the elite class becomes *more* separate—in terms of wealth, power, and lifestyle—from "the other 99 percent," they also become *less* capable of sound decision-making, innovative thinking, and egalitarian justice. Within the ranks of the elite is what Hayes calls "fractal inequality"—essentially competition among power-holders who envy the more powerful; among the wealthy who long to be even more rich; and among the glitterati who desire to sparkle even brighter. This is the compulsion that Hayes sees as truly pathological—the state of mind among the

1 percent to join the .01 percent, and among the .01 percent to be ranked with the .001 percent. A few of the wealthiest and most powerful seem to have some immunity to this kind of pathology, but as a group, those in the meritocracy are in thrall to the possibility of rising ever higher. "As American society grows more elitist," Hayes concludes, "it produces a worse caliber of elites. The successful overachiever can only enjoy the perks of his relatively exalted status long enough to realize there's an entire world of heretofore unseen perks, power, and status that's suddenly come within view and yet remains out of reach."

Here Hayes offers more editorial opinion than is supported by the data. To suggest there is some pathology that drives achievers to chronically desire to achieve more, no matter their past achievements, is probably accurate. To characterize it as some new form of evil is too great a leap. Many of my high achieving friends share a restless and persistent dogma of the "nagging next." No matter how many meetings, surgeries, hours billed, tournaments won, or summits reached, the next word that pops to mind is "next." What am I going to do next? It's an unfailing litany of the next five minutes, weeks, months, and years. It seems to be accompanied by high blood pressure and endless amounts of activity. Those more firmly in "Next's" grasp can even exist on just a few hours of sleep.

Does this make them bad people? While Hayes may think so, I disagree. These are driven achievers. While the Bill Gates of the world enjoy lives that few of us can imagine, the introduction of the Desktop and Windows have added immeasurable productivity to education, business, philanthropy, healthcare, and the list goes on. It's hard to know what drives these rare folks who truly change the entire world and witness it in their lifetimes, but I'm certain that we damn well don't want to interfere with it. Whatever luxuries are afforded to heroes may not motivate heroism, but I think most would be unwilling to take it away from them.

Restoring Our American Dream

So...What Will It Take?

Does this mean there's no hope of what's loosely called "Wall Street reform?" Does it mean that we simply have to give up trying to get these guys to change course and accept whatever economy-busting surprises they deliver next?

No, we are not without hope, we will not damn all for the sins (no matter how egregious) of a few, but we need a clear strategy. Those in power need to be mindful of how they got there. They need to remember not only their own efforts (for which they may be deservedly proud) but remember the rules and protections and opportunities that allowed them to pursue their potential. America and free market Capitalism are a great marriage. Contracts, individual liberties, and property rights are part of the foundation on which millions have built lives and livelihoods.

All around us—from restless Tea Party activism to Occupy Wall Street—there are signs that change is in the wind. We need to restore trust in America's institutions and open up opportunities for a vast majority who seem to have little hope of bettering themselves in this generation.

All well and good. But as a recipe for reform, it doesn't lick a key issue that we hear in that taped conversation between Doug Goldberg and Stewart Wolmark. Let's listen again.

Wolmark: "Hey, congratulations. You got another one."

Goldberg" "Yeah, thank you. Thank you."

Wolmark: "You're hot!"

"I'm hot? Hot with your help, sir."

It would be remarkably decent if we could change the conversation between the two lowly members of the elite to reflect a true awareness of accountability. Something like:

Wolmark: "Hey, I hear you just did a deal on behalf of our company."

Goldberg: "Yes, that's right, sir."

Wolmark: "Well, you know, our firm is ultimately going to be held accountable for the outcome of this deal, so I just want to make sure that the bidding was blind—there was no communication among the participating banks. Because, you know, we want these communities to be getting the highest possible interest rates that banks can afford. Oh, and Goldberg, I also want to make sure that our commission is exactly what we declare it to be when we talk to the municipal representatives. You know I don't want to hear later on that we got something on the side from some other source for rigging the deal. Can you guarantee that?"

Goldberg: "Oh, yes, absolutely."

Wolmark: "Well, that's great. Because we *are* going to be held accountable."

Fantasy dialog?

Indeed.

Morally correct. Ethically correct. Evincing the transformative power of good citizenry.

But pure fantasy.

Why?

Winning

We like to win. Some, among us, love to win. And there are those among us to whom winning is everything.

Try to purge that out of the human equation, and you will be embarking on the most frustrating and silly-minded act of futile social engineering that can be imagined.

Look at Angela Merkel screaming and clapping her hands when the Germans beat the Greeks in soccer. Feel the rush of triumph that Goldberg gets from Wolmark when his boss says, "Congratulations! You're hot!" The euphoria of a lottery contestant informed that she has won millions of dollars. The heart-pounding exuberance of crowds at football or baseball games when the final score is in favor of their teams. The glow of satisfaction on the face of a new homebuyer stepping, for the first time, into a place she can call her own.

To win, the professional football player will run all kinds of risks of severe cranial damage. To win, the formula-one racer will repeatedly put his life in danger. The hedge-fund manager will repeatedly take huge risks. And the poker player will stay in the game long after he should fold.

We not only love the feeling, we love to ogle the winners. The movie stars who win the Oscars and the songsters who win the Grammies. The winners of elections. The gold-medal Olympians. And because money, in our culture, is a significant sign of winning, we gaze on the luxuries and adornments of the rich, compare and contrast their trappings with ours, and imagine what it would be like to win such wealth. We're suckers for this feeling, and there is nothing in the teaching of all religions, the preaching of all ethics, or the practice of all disciplines that can deprive us of the longing for what it feels like to win.

If this is pathology, well, we'd better make room for it because it's not going to go away just by wishing it away. And it's no good lecturing ourselves about the transience of triumph. True, we're not all addicts of winning. But when we have a choice about what to experience, what to admire, what to aspire to—and what to make sacrifices for—winning has, for most of us, an irresistible appeal.

Most of us are just not emotionally equipped to resist the lure of winning. Sure, we know the gain may be short-term. The exultant feeling won't last. But we're drawn to winning like bees to nectar.

Now What?

Do we throw up our hands and say, forget ethics, forget fair play—it's all about winning? Shall we have a free-for-all on the playing field of our society?

Obviously not. And this is where a different kind of responsibility comes in. The responsibility for governance.

In sports, these are called rules and they're enforced by referees and umpires. Throughout the history of civilization, as far as we know, social groups have been quite capable of rule-making for athletic competition without government mandates—a tradition that continues to this day. Sports franchises that involve multibillion dollar investments and attract the attention of rapt fans are run almost entirely by the teams and owners who agree on the rules and regulations of each sport. There are no congressional mandates. The Oscars and Grammies and Tonys are, somehow, decided by those within the industry who establish the rules of eligibility, participation, and voting. Within the realms of academia, the standards set for valedictorians and salutatorians (these are, after all, the "winners" in that venue) are established and enforced by the academic community.

And so it goes. Nearly every professional organization takes responsibility for governance of its own members. Without medical school faculties and boards of review that establish standards for admission into the medical profession, we would have a far more difficult time telling the difference between a licensed medical doctor and a quack. Bar associations designate who and who is not a lawyer, and they do it by governing the standards of admission to this elite club. Nurses and plumbers get licenses. Technicians get diplomas. The tradition of self-governance is as old and respected as the earliest guilds. To maintain one's status and reputation means adhering to the rules and guidelines of the guild.

Which, of course, brings us around to Wall Street. What are the rules there? And wherein does the responsibility lie for *making and enforcing the rules?*

It's an important question, because, without those rules, the financial playing field could become a free-for-all.

Did I say *could become?*

I mean, *has become.*

What could be, and should be, responsibility for the professional rulebook recognized by the most talented financial players in the world has, instead, been forfeited to the one set of umpires that this community trusts the least—the government. It's like handing over the rulebook of professional baseball to the local PTA. As P.J. O'Rourke says, giving power and money to Congress is the same as giving a fifth of whisky and a set of car keys to teenage boys. In the hallowed halls of Congress, senators and representatives with only the sketchiest understanding of the workings of Wall Street are trying to determine the best way to avoid another financial meltdown. Even if their intentions are the best, this would be beyond the scope of their capabilities. (And, as we know, many of their intentions are far from the best: they are simply taking steps to insure their own reelection.) Those in positions of responsibility within underfunded government bureaucracies concerned with oversight are grappling to enforce regulations that range from the challenging (cases of insider trading) to the downright impenetrable (requirements for bank reserves).

The point here is that despite their best intentions, members of Congress are regular folks. They are lawyers and business people, educators and community organizers. But they are not economists or investment bankers. They don't have the expertise to understand the hugely complex issues that threaten global finance. Unfortunately, they all seem to possess an acute barometer of public opinion, which, in the absence of expertise, must suffice.

On the great stage of public concern, we are treated ever more frequently to the sight of Wall Street executives being hauled before Congress and grilled about the most recent scare or disaster (multibillion-dollar trading losses, use of customer accounts for high-risk gambits), resulting in an outcry for a new set of impossible-to-follow and impossible-to-enforce regulations. Every once in a while, the FBI or SEC can haul malefactors into court where, after the expenditure of millions of hours in lengthy examination of the skunk-trails left by emails and many millions of dollars in legal fees, the rule-breakers are as likely to come away with a hand-slap fine (and "no admission of wrongdoing") or a few years of time in low-security prisons (once again at government expense). Then, when the most recent stage play between the members of the finance industry and government regulators has been acted out for the public, leading members of the financial community will repeat their outcry against "government oversight" and "government regulation."

When did the most sophisticated financial system in the world turn into a game of cat-and-mouse between the kids on Wall Street, the politicians in Washington, and the government watchmen with their billy clubs?

Time for this game to stop. Wall Street needs to grow up, take responsibility for Wall Street. Sure, it's like herding tigers to get these guys and gals together, talking about what's to be allowed (and what isn't) on this playing field. Every one of these players is out to win, and win big, in the most cutthroat game on earth—the rivalry for big bucks. But if this league of heavy hitters can't take responsibility for their own actions, there's no way they're going to be forced to look at the consequences of their actions. And unless they do that, there's no way Wall Street is going to deserve to win the public trust.

My long-time friend J. Carter Beese would say that U.S. markets commanded a "transparency premium." FINRA, the SEC, GAAP accounting rules and our legal system each work to ensure that investors in our markets can be confident they are getting a fair deal. Carter argued that that perception and trust were worth a lot and provided empirical justification for ongoing vigilance.

Chapter 9

Earning Trust
Economic Bedrock

At some point I decided it might be a good idea to show viewers of *Wall Street Week* what a million dollars really looked like. So I went to the place where you'd expect to find a million dollars. A bank. Not the biggest bank in North America, but not a small one either. Just the kind of branch facility you find on street corners of most American towns and cities.

When I explained what I wanted to do, the bank manager was happy to comply. He would need to ensure proper security, of course, before the cameraman could come on the premises and film the stack of bills. That wasn't a problem. What was a problem, the manager said, was how to get together a million dollars.

I asked how long it would take him.

He wasn't sure, but he would get in touch with me as soon as he had it.

A couple of days passed. Then a few more. Finally, I got the call I was waiting for. It had taken a week and a half, but the bank was ready. The manager had a million dollars.

What many people don't *get* (and by "many people," I include some very affluent and intelligent business executives) is the simple fact that the bank does not have your money. This fact is explained in most Economics 101 textbooks. What happens is that trusting depositors pour money into a bank like milk into a pitcher, and then the pitcher pours it out into little cups of interest-producing investments around the community, the country, or the world. What apparently doesn't sink in—judging from the casual conversations I've had from time to time—is that the money flowing *out* of the pitcher far exceeds the money flowing *in*. So *your* money—from your paycheck and your savings and Uncle Todd's birthday gift—isn't there. The bank I visited, for instance, probably got many millions of dollars in deposits from trusting clients the day before, but most of it was not cash, most were checks and electronic transfers. The velocity of departure from the pitcher was so great that when I showed up and asked to photograph a million dollars, the money was already gone, or more to the point, had never physically been there at all

And yet I wouldn't be surprised if you make use of a bank to keep your personal funds and dole them out when you need them.

Why do you do that if you know (from that Economics 101 textbook) that the money you put into the bank is, almost instantly, gone?

The reason, of course, is trust.

Most of us are using banks. Many of us are trusting in them. And yet, around June of 2011, three years after the 2008 crash, Gallup Polls showed that Americans' confidence in banks hovered around 23 percent.

So here's the question. *Why on earth would you put your money into an institution you don't trust?*

Levels of Trust

Earlier in this book I described what I believe happens when trust in institutions begins to erode. In this chapter, I want to take the next step, toward determining what it takes to *earn* trust—a key step, if we are to have a strategy of hope in this society. But before we can get to the topic of earning, we have to clarify what kind of trust we're talking about. Because, as shown so clearly by the example I just gave, there can be a huge difference between what people say (in surveys) and what they actually do (in real life).

While economists agree that trust is important, even critical, to the functioning of social units, there are actually a variety of concrete actions and common pursuits that require unusually big gobs of social trust. We take it for granted, for instance, that Americans are going to drive on the right side of the road. A trivial example, perhaps, until you speculate what would happen on the roads if there were no common agreement recognized by all drivers at all times of the day and night. (Well, actually, we do know what happens—tragedy—when a driver declines to obey this most basic form of social behavior.) We also have multiple examples of trust in the form of decisions we make in microseconds throughout the day. During your time in the grocery store, as you go aisle-to-aisle filling your cart, the manager and all the checkout people give you the latitude to roam freely, collecting the store's goods, for the simple reason they trust you to pay for all those goods before you leave the premises.

And so it goes. Most of us trust that we'll have a steady flow of water and electricity to our homes if we make the appropriate arrangements with suppliers and pay our bills. We trust that traffic will stop at a red light and proceed on green. We trust that, most of the time, most of the people around us are going to restrain whatever murderous or larcenous tendencies they harbor in their souls, allowing us to remain mostly unscathed as we go about our daily lives.

Let's call this *operational trust*. And even though it may seem humdrum, it should not be underestimated as an important component of social trust. Proof? Well, just consider what would happen in a supermarket where people are frantically filling up their carts with no intention to pay. We've seen those conditions more than once. They're called riots. And, as we know, they are the direct opposite of social order. Or try living in a city such as post-occupation Baghdad or post-earthquake Port-au-Prince where electricity goes on and off at unpredictable intervals, and you will become acutely aware of how much operational trust in civil amenities assists in the maintenance of bearable living conditions and the reliable conducting of everyday affairs. We are fortunate to live in a society where, most of the time, skyscrapers don't fall over, planes rarely plummet from the sky, and brakes, when applied, usually perform marvelously. What we call "taking things for granted" is actually comes from a long habit of building up operational trust in things that regularly serve us, as a society, and in their own methodical and unglamorous ways provide us with a satisfying degree of social good.

While I don't know for sure that any Gallup Polls have been conducted to find out our level of operational trust, my guess is, the percentages are pretty high. Oncoming cars keep passing on the left. Buildings remain in upright positions. Goods are paid for before customers leave the store. Turning on a light switch, almost invariably, produces illumination. We may have gotten a bit smug about our dependence on these trustworthy occurrences, but they really do help to hold the fabric of society together.

Well, then, why has the lack of trust in various *institutions* in our society become so pervasive? Why was it that in midyear 2011—at the same time the trust in banks fell to 23 percent—only 34 percent of Americans had any trust in the public schools; 28 percent trusted newspapers; 21 percent trusted the unions; 19 percent trusted big business; and 12 percent trusted Congress? How can we possibly begin to restore *institutional trust* (as differentiated from operational trust) when we have statistics like that to start with?

Perceptual Evolution

When we wonder how to restore trust in the most fundamental institutions of American enterprise and democracy, we are really asking two questions – How can these institutions can be perceived as trustworthy, and how can they sustain the level of trust that is needed to continue serving the public interest?

Ideally, of course, we all could proceed with a high degree of confidence in all the business, military, political, religious, media, and financial institutions in America. Such bliss is not likely to occur either now or in the future, for a variety of reasons. Perception plays a huge part. When sex scandals rock the church, public confidence in churches tends to decline. When the economy tanks during a presidential term or a congressional session, confidence in the president and Congress inevitably drifts downward. When we're repeatedly told that standards of public education are declining and lower percentages of students are graduating, trust in public schools ebbs sharply. And when the media itself comes under suspicion as being too biased, too lazy, or too entertainment-oriented, public confidence in TV and newspapers takes a hit.

All these perceptual issues have minimal impact on our day-to-day lives. We sigh, bemoan, wish that institutions would be better, more reliable, that politicians wouldn't be so grasping, execs wouldn't be so acquisitive, that kids would listen to their teachers and newscasters wouldn't be so windy. It's always disappointing when our expectations aren't fulfilled and when we can't trust people who, we feel, ought to behave better.

But there's one set of institutions that absolutely requires our sustained trust and confidence for the day-to-day operation of our economy. I mentioned them up top. They're called banks.

Banking on the Future

To see why trust is so important to the banking system, let's revisit that bank where I asked to see a million dollars all in one pile. As it happens, I know enough about banks—and remember enough from my economics textbook—to not be surprised when it took eight days for the manager to gather enough stacks and bundles of cash to comprise a cool million. Since it was no surprise, I did not panic when he could not immediately produce the cash—nor did I run next door and tell my friends that the bank was strapped for money. Having faith in the manager's good word, I waited until he got all the greenbacks together, then went in with the cameraman to get our visual of a million-dollar bounty.

Of course you can see where I'm going with this, because history has taught us exactly what *does* happen if I and my neighbors and all my neighbors' neighbors simultaneously descend on all the banking institutions in the country and demand to see our money immediately. If we put it there, we trusted them, and if they can't cough it up, then where has the money gone?

This, as we know, is called a run on the banks, and it's just about the worst thing that can happen in any economy. (What we saw in 2008 was also a run on the banks—but in a different form and at a higher level. In that case, the banks were in such a hurry to collect *from each other* that few, if any, had the necessary billions in assets on hand to meet their multibillions they had outstanding in promissory notes to other banks. But I digress.) And we also know the secret perceptual component that makes all the difference between a banking system that remains operational and one that goes to hell in a handbag. It's called "confidence."

Now you see it, now you don't. If we're all confident that we can drop by our respective banks any time and withdraw the money we've put in (maybe with a tad of interest here and there), then, hey, what's the rush? Leave it there when you don't need it, take it out when you do. But when confidence goes kaput all at once—if I'm telling my neighbor the bank is going broke, and you're telling your neighbor the same thing and we're all rushing to our ATMs at the same time—it's curtains for the system. Because we're all going to get the stern reminder that most of the money we put there is no longer there. It's all been poured out into cups and bowls and saucers distributed elsewhere. That herd action led to disaster in the 1930 crash. And again in the 2008 crash. And will, again, whenever confidence in the banks collapses.

(Of course, today we have some insurance against such cataclysms, in the form of FDIC insurance on personal accounts in private banks and in the form of TARP monies used to bolster investment banks. Thereby, we forestall instantaneous crashes that could occur. But even with this insurance in place, total bank obligations far exceed any amount they are actually capable of paying.)

Hence, the conundrum. As long as people have trust in their banks and continue to deposit money without fear of loss, the more the banks can be trusted. But loss of trust, when it happens, becomes a landslide that gathers momentum as it goes downhill. As demonstrated in the 30s, a "run on the banks" forces the banks to close down. The closing down of banks creates further panic—causing still more depositors to rush to the closing doors and hammer away, hoping for the impossible. They can't get their money back for the very simple reason that the money is not there. But the money was never there! The only thing that changed was public perception—from seeing the banks as reliable, trustworthy depositories to a perception that the banks are the sanctuaries of thieves and scoundrels and ruthless exploiters. In a flash they can go from being expediters of the common good to the representatives of universal villainy. Which is what happens when trust in the banking system collapses.

The Guy Who Saw This

Thanks to the rigorous study, political persistence, and sheer bull-headedness of an undocumented immigrant from the little island of Nevis in the West Indies, we have one of the most innovative financial systems in the world. It's all built on confidence. But he planned it that way.

When Alexander Hamilton became first Secretary of the Treasury on September 11, 1789, he hired the largest staff in the nascent government bureaucracy—thirty-nine employees (compared to five in the entire State department). He set them all to work on a variety of accounting and clerical tasks in a concerted effort to straighten out the books of the newly conceived and constitutionally created central government. Hamilton took it upon himself to formulate a banking and monetary system that would serve all thirteen colonies—and whatever states were to be added to the Republic—in all the years ahead. It was a daunting task, for sure, and he would later get skewered by many rivals for taking such a high-handed approach. But as he saw it, there was a task that needed to be done, and done soon, if the neo-United States was going to thrive as a single unit and become an economic powerhouse competing with the hegemony of Great Britain and Europe.

First problem, the currency. A hodgepodge of coin and specie had come into circulation during the Revolutionary War. Before Hamilton's term as U.S. Treasurer, when he created the first private bank in New York (in 1784), it was partly with the intention of helping to make order out of monetary chaos. As Ron Chernow describes the situation in the biography *Alexander Hamilton*, "By the end of the Revolution, it took $167 in continental dollars to buy one dollar's worth of gold and silver. This worthless currency had been superseded by new paper currency, but the states also issued bills, and large batches of New Jersey and Pennsylvania paper swamped Manhattan. Shopkeepers had to be veritable mathematical wizards to figure out the fluctuating values of the varied bills and coins in circulation."

To perform the equivalent service for a centralized United States government, Secretary Hamilton surmised that he would not only have to create a mint where a universally recognized currency could be produced, he would also need a Treasury Department that would perform a variety of functions serving as a repository for excise taxes, extending credit, and collecting customs duties. To facilitate all this he submitted a massive proposal (140,000 words), his famous *First Report on Public Credit*, which described how the Treasury would assume all the debts of all the states as well as the combined outstanding debt of the U.S. government ($54 million in 1790, the equivalent of more than $4 trillion today, measured by percentage of GDP.)

How to pay for all this? Income would come from post office revenues and later, as proposed by Hamilton, from customs duties levied by the Federal government (rather than individually by the states) and from excise taxes on "luxury goods" like wine, liquor, tea, and tobacco. Bonds would be issued by the Treasury at an attractive fixed rate of interest, and the bank would buy back or retire about 5 percent of its debt every year. Hamilton foresaw the stabilizing effect of a single bank dealing in a single currency. If all went well, government bonds (initially called "scrip") would function like loan collateral, thereby effectively increasing the amount of available credit in the economy. To manage the government debt, revenues would be set aside "at regular intervals" to pay the interest on the bonds that had been issued and pay back a portion of the principal. Hamilton hoped there would always be enough trust in the government bonds to ensure that investors did not indulge in wild speculation. As Chernow points out, "*Hamilton intuited that public relations and confidence building were to be the special burdens of every future treasury secretary.*" Or, in Hamilton's own words, "*In nothing are appearances of greater moment than in whatever regards credit. Opinion is the soul of it and this is affected by appearances as well as realities.*" [My italics!]

The adoption of Hamilton's plan would, he hoped, contribute to a boom in U.S. commerce and manufacturing by increasing, across the board, the amount of credit available to the public. True, the system demanded oversight to ensure that revenues were sufficient to fund interest obligations. If there was going to be substantial, reliable ongoing public credit, Hamilton asserted, "The creation of debt should always be accompanied with the means of extinguishment."

Despite objections from Jefferson, Madison, Adams, and others, Hamilton's plan was passed by the Senate and House and signed into law by George Washington. In one respect, the creation of the U.S. Treasury, combined with establishment of a single U.S. mint, had the desired effect. Credit increased, business prospered, and there was a surge of expansion in trade and manufacturing, especially in the north. But there was also blowback. The imposition of taxes (needed for "extinguishment" of the debt) led to charges that the U.S. government was no better than its forerunner in the British monarchy—and, before long, led to violent outbreaks of protest such as the Whiskey Rebellion. The other problem was speculation in government securities, which began to take off from the moment of the first announcement of the new policies. As Chernow observes, "Hamilton's *Report on Public Credit* had an electrifying effect. Securities began to change hands with a speed never before seen in America."

Hamilton further bolstered confidence when, in early 1791, his bill to charter the first central Bank of the United States was passed by the Senate and then the House. By then, government scrip was being bought and sold at such a furious rate that prices were skyrocketing. Hamilton publicly deplored these "extravagant sallies of speculation" which, he said, "do injury to the government and to the whole system of public credit," but words alone were not enough to stem the tide. Within a month, bank scrip that had been sold for $25 was changing hands among leading speculators at twelve times that price. Then, on August 11, 1791, the bubble burst and prices crashed.[58] Clearly, appearance had outstripped reality, and in the first recorded freefall of American securities, prices fell back to earth. Despite Hamilton's efforts to squelch speculative forays, the market had played its hand. "The superstructure of credit is now too vast for the foundation," he observed. "It must be gradually brought within more reasonable dimensions or it will tumble."

By the time Hamilton spoke these words, three new banks had been opened in Manhattan, and speculation in both government and bank securities was, once again, reaching new heights. In early 1792, share prices again began a sharp decline and Hamilton's Treasury department began a buyback of securities to stabilize the market. "Instinctively," Chernow notes, "Hamilton understood the creative ambiguity necessary for a central banker coping with a crisis…He restored temporary calm to the marketplace."

Soon after, a remarkable thing happened. Two dozen brokers, unaffiliated with the government, with no mandate but their own, met under a buttonwood tree at 68 Wall Street with a single-minded purpose—to establish rules that would govern securities trading.

The Rules of Play Today

Now, more than seven generations later, we have had a decent interval in which to assess the consequences of the first Treasury Secretary's brainstorms as well as the evolution of the broker group that met under the buttonwood tree.

The First Central Bank had a charter of twenty years. Though the charter was not renewed—due to intense political opposition—

a second and then a Third Bank were established during the 1800s. Finally, 1913 saw the creation of the Federal Reserve System with its twelve regional banks overseen by a board of governors appointed by the President and confirmed by the Senate. The Fed is designed to be politically independent, but is answerable to Congress (hence, the hearings during which the current Fed chairman appears before the Banking committee and gets thoroughly grilled). Banks in the Federal Reserve are the intermediaries between the U.S. Treasury and private banks, and the Fed is considered the "lender of last resort" for the entire U.S. banking system. In practice, this means that the U.S. Treasury has its own account within the Federal Reserve System where tax dollars are deposited and from which payments are dispensed. The Fed also serves as a conduit for putting coins and paper currency into circulation, and it sells and redeems U.S. government securities. Its broadest function—and still the most closely scrutinized—is to expand and contract the money supply as economic conditions change. This it does largely through adjustment of various lending rates offered to private banks, giving them opportunity to borrow on favorable terms during times when credit and cash are needed to meet the demands of customers. In addition, the Fed has regulatory power over banking institutions, and it can take steps to protect the credit rights of consumers.

Thus, at the end of some two hundred years of political battles and technical adjustments, the role of the central bank is in many ways just as Hamilton conceived it. The Board of Governors arrives at a single monetary policy that is enforced by all twelve regional Federal Bank branches. Expansion and contraction of the money supply can be adjusted, to a degree, by implementation of Fed policies. Theoretically, then, this incarnation of the central bank should provide a buffer to counteract market fluctuations (boom and bust) that have proved to be endemic to the functioning of an open economy.

But what became of that small community of brokers—entirely independent of any government operation—who met under the buttonwood tree on Wall Street, the group that would evolve into the New York Stock Exchange?

Love it or hate it, this is our heritage. We have a centralized banking system wherein the interests of every one of us – from the purchaser of Altoids at the corner newsstand to the hedge fund manager in the global network—relies on a U.S. Treasury and Federal Reserve System to back up the validity of the dollars we exchange with each other. We use private banks, credit unions, and other financial entities whose operations and profitability are closely linked to and regulated by the public institutions that affect their (and our) well-being. The reality of working with those institutions is that we all draw on a vast pool of trust. We may or may not always love the way the system works, but in our daily lives we rely on banks to produce the money that we deposited there (even though the money isn't there) while the banks, in turn, rely on borrowers to pay them back (even though the borrowers don't have that money today) and on a central bank to lend them money, literally overnight, when they need it.

As for that group of brokers who met under the buttonwood tree—and their many incarnations in succeeding generations—anyone who invests puts some of our trust in them, too. We expect them to act in their own interests and in the interest of their clients, as they did in Hamilton's day when they set rules that put an end to speculative frenzy. We rely on them to set the boundaries that will apply to anyone who participates in their commercial ventures. And we expect them to duly referee what would otherwise be an anarchic game with no boundaries.

This is the sum of the trust we place in our government and in Wall Street.

Which is what makes the collapse of 2008 particularly awful.

In this case, the run on the banks was not by citizens pressing against locked doors and clamoring to get their deposits out. As noted, the run on the banks was by other banks. As is always the case with banks, the money wasn't there. It had never been. But what collapsed was trust. One fine morning, one bank woke up without confidence in the next. When they rushed to the doors and requested their money, it wasn't there. And so that bank turned to the next…which, in turn, went to its neighboring bank. And when the run on the banks was over, it was all of us who had to step in—in the form of all the good faith and hard work that has built the U.S. Treasury—and rescue the banks from their folly.

Those brokers under the buttonwood tree failed to keep the trust. We never asked or expected them to guarantee that we would win. All we asked was they set the rules for and declare boundaries of fair play in their game. They didn't. And when the time came to blow the whistle, they had no refs in place.

So here we are with the aftermath. A damaged economy. Huge debt. Empty houses. High unemployment. Ruined expectations. But beyond and above all that, a deep distrust about the institutions we usually consider so trustworthy. This time, they failed us.

Getting It Together

Rehabilitation won't be done with a great PR campaign and a wave of the flag. Too many of our assumptions about our banking system and Wall Street have been shattered. Getting back our hope for the future depends on re-establishing trust in the institutions that so recently failed us.

Will rules be broken? Will mistakes be made? Of course. But as Hamilton himself observed, there is a positive side—both in terms of private gain and public good—that far outweighs the negative. "If banks," he wrote, "in spite of every precaution, are sometimes betrayed into giving a false credit to the persons described, they more frequently enable honest and industrious men of small or perhaps of no capital to undertake and prosecute business with advantage to themselves and to the community."

Some two hundred years after Hamilton wrote those words, this notion of how to use our capital is deeply woven into the American DNA. It is no coincidence that one of the best-loved and most tenderhearted American films of the twentieth century contains a message of aspiration that is almost an exact duplicate of Hamilton's assertion. In "It's a Wonderful Life," the courageous young man played by Jimmy Stewart takes a stand on behalf of his fellow townspeople, speaking against those who have betrayed their trust. The time period is the Depression, when bankers around the nation were denying credit and foreclosing on property of their fellow citizens. And George Bailey, played by Jimmy Stewart, faces off with the grasping Mr. Potter:

Now, hold on, Mr. Potter. You're right when you say my father was no businessman. I know that. Why he ever started this cheap, penny-ante Building and Loan, I'll never know. But neither you nor anyone else can say anything against his character…why, in the 25 years since he and his brother, Uncle Billy, started this thing, he never once thought of himself. He didn't save enough money to send Harry away to college, let alone me. But he did help a few people get out of your slums, Mr. Potter, and what's wrong with that? Doesn't it make them better citizens? Doesn't it make them better customers? Just remember this, Mr. Potter, that this rabble you're talking about… they do most of the working and paying and living and dying in this community. Well, is it too much to have them work and pay and live and die in a couple of decent rooms and a bath? Anyway, my father didn't think so. People were human beings to him.

And, George might have added, human beings need to dream.

Chapter 10

Going for the Dream
Hope, Trust, Striving, Yearning, Determination and Grit

From every corner of the world we came. By ship, overland, and through the air. Some of us had been forced to leave; some had fled; some came as slaves; some as passengers. We were smart and stupid, rich and poor, close to starvation or flush with wealth or merely wandering and lost. Some had children; many were single; still others pioneering for friends and family left behind. What a strange, diverse crowd—and not always the most savory of characters.

Once here, none of us found what we were looking for, because none of us quite knew what we had expected to find. For some, the rumors of great opportunity turned out to be true enough. Others found that the truth had been suppressed—due to secretiveness, shame, or despair—by those who arrived earlier. There were fetid accommodations, crime, unemployment. Prejudice and misunderstanding prevailed among new neighbors. Rivalries were renewed and feuds reignited among old enemies. Even in America, people could be enslaved, children could starve, disease could run rampant, larceny and deceit could triumph while industriousness, intrepidness, and creativity suffered. Profits sometimes turned to plunder while indigence could become the endgame as service and self-sacrifice were punished. It's understandable that we smile when

Yogi Berra says, "If you don't know where you are going, you'll end up someplace else."

So why did we stick it out?

We—and by "we" I mean, of course, your folks and mine—had no real strategy. We only had hope. So hope became the strategy.

And it still is.

Have you read the stories and seen the images of American cities scarred by hurricanes? Among the acres of homes that were flooded and destroyed, in the middle of chaos, did you really think someone would come up with a single ingenious strategy for making repairs, restoring services, repaying the costs, and restarting broken lives?

The same questions could be asked when the Twin Towers fell, when the financial markets nearly came to a standstill, when terror of a viral epidemic seized the country. We've held on to hope through earthquakes, drought, floods, wars, assassinations, recessions, disease, depressions, riots, rebellions, and terrorist acts.

Not bad for a motley group of strangers gathered from around the world who somehow make it all work.

Still, we could miss the boat if we don't remember our main strategy: *Hope, trust, striving, yearning, determination, and grit are what it's all about.*

Way back in Chapter 3, I introduced the musings of psychologist C. R. Snyder, who proposed that three things are needed to make the "journey of hope." Those key elements he defined as goals, willpower, and waypower.

For the sake of argument (and I'm sure there will be those who argue with me), I think we Americans shine when it comes to willpower and waypower. Just measuring in terms of productivity (a very crude but important measure of labor output), we rank highly if not highest among the leading nations of the world. In all the examples I have just mentioned—from Hurricanes Katrina and Sandy to the attacks of 9/11 and financial mini-Armageddon—we have come roaring back with brio, engaging in public debate, vigorous action, and stupendous shows of initiative. Despite the sometime rancorous tensions between government and private institutions and their respective representatives and allies, we have somehow managed in all major instances of this kind to coordinate the actions of our governing bodies with the capabilities of private enterprise and the needs of individuals in order to create *not* a perfect union but certainly a united effort to get things done. As doers, despite all our disagreements, we tend to excel.

So much for willpower and waypower. But what about that third element—goals? Here, I'm not so sure. And in order to get to the essence of the strategy of hope, that's how I would like to end this book—with a look at our goals.

Shared Goals

The Dutch have a big problem.

In Holland, about two-thirds of the land would be covered by seawater were it not for a system of dams, dikes, and massive gates designed to hold back flood waters. This long-term project, which involves the building and rebuilding of dikes, the control of river floods, and the pumping of water from one place to another, has been going on continuously for about eight hundred years. As an example of social cooperation undertaken steadily over the centuries, this one is hard to beat. Of course, there was always an urgency to the task at hand. Let the dikes deteriorate over time because of community indifference or lack of oversight, and it won't be long before your own field as well as your neighbor's is drowned in an overflow of chilly and brackish North Sea. Clearly, all could see the threat to agriculture, community, and housing, and all were strongly incentivized to keep their dikes in good repair and anticipate, as much as possible, the negative effect of storms thrashing against seawalls.

Eight hundred years is a long time to maintain consistent public support for a public works project, as anyone who has ever attended a few contentious zoning board hearings will attest. No one really believed a breach in the dike could be plugged by a boy standing on tiptoe. What was required was incessant construction, reconstruction, reinforcement, and innovation. During the Middle Ages, the front side of each *wierdijken* or earth dike would be packed with seaweed held in place with poles. (As the seaweed compressed and rotted, it formed a dense protective outer coating.) For a few centuries, dikes were fronted with regularly maintained reeds or wicker mats, but eventually these were swapped out for timbers. When invading shipworms began to wreak havoc (around 1730), the Dutch dike preservers began investing in stone, imported at great expense. Today's dikes are sand covered with clay, often protected with crushed rock, basalt, or a layer of paving. Up top, along the ridges, sheep graze on the grassy plateau, efficiently tamping down the soil with their hooves while their mandibles manage grass control.

Obviously, this is not one of those publicity-grabbing short-term projects so beloved by elected officials seeking favor from their constituencies, nor is it the kind of corporate enterprise showing quarterly returns that bring ear-to-ear grins on shareholders' faces. One would think that either an eighth-century dictatorship or a powerful monarchy would be required for such a civil engineering project. And yet the Dutch seem to have accomplished all this without such extreme forms of top-down management. (There is indeed a royal family, but dikes are far from their leading concern.)

What, then, keeps the dikes in place? Who is responsible for the maintenance contract? Who can they blame, and fire, when floods occur (which they do), and what must be done to meet the challenges of an ever-rising sea-level?

Since Dutch farmers had the most to gain from water control, they were the first to get involved, cooperating to form nongovernmental water boards to look after the dikes adjacent to their property. Every three years, a director from the water control board visited, making sure each farmer was doing his bit. But what about the folks farther inland who also reaped benefits without lifting a finger to help? By 1798, a public works administration (the *Rijkswaterstaat*) was set up to oversee and support the work of the individual water control boards. To this day, the 27 water control boards (down from some 2700 in the mid 1900s) hold their own elections and levy their own taxes.

If ever there was a shared purpose, this is certainly exemplary. It is very unlikely that a twelfth-century monk, even a very clever and prescient one, could have planned ahead for the elaborate phalanx of modern dams and drainage systems that, today, help control the surge of the North Sea. With a massive set of floodgates serving as a storm surge barrier, the current system is continually being updated to meet the kinds of conditions that occur about once every ten thousand years while an elaborate warning system has been set up to coordinate evacuations, barrier closings, and dike patrols.

There are no guarantees. (The foe is, after all, nature with its many wraths, now accompanied by human-assisted global warming.) It is, however, a goal shared by all—that the waters will be controlled, disasters averted, and a nation and its people preserved. Some eight centuries after launching this project, it's still going strong.

In America we have a much wider range of natural disasters. Obviously, the Dutch model of shared interest does not apply to our environmental concerns. And yet we do have shared a common goal, hammered into shape by a group of diligent debaters who conceptualized a common purpose—in their words, "to form a more perfect Union, establish Justice, insure domestic Tranquility, provide for the common defence, promote the general Welfare, and secure the Blessings of Liberty to ourselves and our Posterity." For a couple of hundred years, we've been chasing this objective. We still have a long way to go. But if we're going to reach that goal, we need to keep reinventing our storm surge protection.

Floodgates

We have just come close to a financial meltdown. We've patched up some of the damage, cleared out some of the wreckage, read the riot act to the head-bashers.

But what is our goal here? What innovations are needed to create a long-term operational system that incorporates current technology, remains serviceable on a worldwide platform of mutually dependent operatives, delivers capital investment and liquidity for a healthy economy, and serves the interest we all should share in protecting ourselves from systemic failure?

A tall order, I know. But if we don't work toward that goal, what do we have to hope for? Another emergency—two or five or ten years out—when, again, the denizens of Wall Street will flee to the harbor of the Capitol to confer, at length, about how to patch the dikes?

The last time it happened, there were plenty of people saying, "Well, now we know what needs to be done so it never happens again." But what does need to be done? Who has a blueprint they'd like to show us?

These are *our* fields. Dike maintenance is *our* business. If we wait passively for some all-seeing financial management director to come along in the form of a president, Fed director, or congressional committee to design a plan, we'll wait forever. This is not a top-down management thing. The dikes are weak, and if we're going to maintain as well as repair them, we'll need the cooperation of Wall Street as well as Main Street, of business as well as municipalities, of taxpayers as well as dependents.

Fortunately for us, Americans are very good at this kind of thing.

In June of 2012, four years after the great financial tremor, my colleague Douglas A. Kass gave a very interesting commencement address to the graduating class of Alfred University, where he had spent his years as an undergraduate. (Later, he would go on to get his MBA at the Wharton School at the University of Pennsylvania.) "Interesting" is the word I use, for a number of different reasons. Doug has a background as a Nader's Raider (lefty). His current position is a hedge fund manager at Seabreeze Trading (well up in the upper one-percent). And his commencement talk had to address students with many different political leanings and practical ambitions. That's what I consider interesting.

Over on the left side of his speech, he praised America's tradition, exhibited during the previous two and a half centuries, of thriving through thick and thin. "We have always moved," he said, "toward 'a more perfect union' and back toward stability and ultimately growth." This, he noted, often requires the help of government. Going way back, he mentioned how Abe Lincoln set up the first intercontinental railroad system and the first land grant colleges. He mentioned the "rules and safeguards" implemented after the Depression, to insure that a bust like that wouldn't happen again. Doug cited FDR's New Deal programs, Kennedy's founding of the Peace Corps and helping to end racial discrimination, and Johnson's "Great Society" and "War on Poverty" programs.

Later, putting on his hat as a card-carrying one-percenter, he delivered a hedge-style opinion piece about the state of the non-too-robust U.S. economy, with a glance at the outlook internationally. The American consumer, he said, still remained hobbled, "burdened by wage inflation and weak income growth, structurally high unemployment, lower housing wealth, elevated debt loads and still-tight credit." He fretted about the residential real estate market and the U.S. propensity for relying on 'the kindness of strangers" (e.g., foreign holders of our debt). Other problems: The financial disarray in state and local government, and the growing schism between cash-rich large corporations and small-business and cash-strapped consumers.

But it was in the middle of his speech that Doug became most eloquent in his observations of what has gone awry and what must be done to fix our economic and financial system.

"Wall Street was at the epicenter of all that went wrong in our economy over the past three years. A small cabal of bankers who created unwieldy, unregulated, and unnecessary derivative products, or financial weapons of mass destruction, ended up producing an economic, financial and credit disequilibrium that affected nearly everyone in the audience today. This was done under the not-so-watchful eyes of regulatory agencies and of our government—America rushed headlong into the twenty-first century without a proper understanding of what economic policies and financial tools were going to be required to prosper in a changing world. For more than two decades, the U.S. economy favored financial speculation over production."

And then, speaking to the newly minted Alfred University graduates, Doug said:

"We all won't forget the last few years, but I remain hopeful—and you should be, too—that the safeguards now being put in place will protect us in the future. But…to be totally effective, you must contribute to the debate that dictates policy."

Back to Accountability

Can we hold ourselves accountable for what just happened—and, at the same time, pass the baton of responsibility to the next generation?

If we're going to maintain the dikes, I think we must.

Make no mistake: whether left, right, or middle, we all messed this one up. Wall Street, as Doug admitted to those kids at Alfred, screwed up royally, creating "financial weapons of mass destruction." Government blew it—yes, both Republican and Democratic administrations—throwing out the safeguards installed after the Depression, changing the rulebook to suit the hubris of the players, and totally neglecting its duties as referee. And hungry consumers went for all the goods they could get: like starved foxes in chicken-yard heaven, we've chowed down on everything that wears feathers.

We need new dikes. But even more than that, we need to maintain the ones we have, and plan the floodgates for that once-in-ten-thousand year event.

Do we have the wherewithal to do it? In March of 2012, a Rasmussen survey showed that 63 percent of likely U.S. voters voiced the opinion that U.S. society "is generally fair and decent."

A 2011 AP-National Constitution Center poll reflected the views of 74 percent of Americans who agreed "the United States Constitution is an enduring document that remains relevant today" and a similar percentage (73 percent) approved of the statement, "People should have the right to say what they believe even if they take positions that seem deeply offensive to most people." Despite all the splits between left and right, conservative and liberal, rich and middle-class and poor, 69 percent of Americans surveyed in 2012 were united in their view that, "As Americans, we can solve our problems and get what we want."

We have reserves of trust in our Constitution, in each other, in our rights of free speech, and the confidence we share in our ability to solve our problems (maintain our dikes). These are what people in the economic-studies racket might call "bankable social assets," and they're worth investing in. Some of those graduates at Alfred University—and other universities around the country—will, I hope, contribute to the debate that dictates policy. And, further, I hope they will be fair-minded as they help set policy. I hope they will hold themselves and their peers accountable for what happens as technology moves ahead with lightning speed and the opportunities for empathy become ever more remote. And if they do get in the henhouse, I hope they don't behave like foxes.

If this begins to sound like a public service announcement, so be it. We have lived for a couple of decades with the "greed is good" motto, and it really hasn't done us much good. An even more bitter pill to swallow, it hasn't even made us *feel* good. As reported in the *World Happiness Report* edited by John Helliwell, Richard Layard and Jeffrey Sachs of Columbia University, "In the U.S. ...uncertainties and anxieties are high, social and economic inequalities have widened considerably, social trust is in decline, and confidence in government is at an all-time low. Perhaps for these reasons, life satisfaction has remained nearly constant during decades of rising Gross National Product (GNP) per capita."

As it happens, research shows that once our basic needs are addressed, most of us get very little gratification from an additional injection of cash. What gives us a charge—chemically, mentally, emotionally, and biologically—is our connection with others. Right now, the urgency of connecting is greater than ever before, because we are consuming the world's goods and resources at an unprecedented rate in order to feed the hunger that cannot be satiated by means of consumption. Cooperation in a shared enterprise—fixing the dikes—will not only result in better ways to weather the economic storms, it may present us with the surprise of learning that we are happier doing it than not. *If*, that is, we can work together.

Farr's Four-Step Recipe for Restoring the American Dream

The America I see around me is certainly a melting pot. And one of the challenges of creating a pot is finding the right recipe, which we have been trying to do for several hundred years. Needless to say, this book would be unfinished if I didn't add my own recipe to this joy of cooking—fully understanding there are no perfect solutions to our current problems, but also with the conviction they must be addressed if we're going to get out of some deep trouble that we're in.

The ingredients in the recipe I propose are really quite simple:
1. Reconnect culpability and consequence
2. Enforce existing regulations
3. Celebrate success
4. Remember our American Dream

1. Reconnect Culpability and Consequence

Simply put, if you do the crime, you do the time. When consequence is borne by others, the perpetrators have no reason not to blithely perpetrate again.

As you may have recognized from my earlier comments in this book, I am no fan of social engineering or the automatic redistribution of goods and income assuring that everyone will get an equal share. I recall a Socialist Republic that tried to do such things, a multigenerational experiment that was not only a totalitarian abomination but also, economically, a colossal failure. What it left out of the social-engineering equation were the driving forces of human nature that I have often mentioned in this book—hope, trust, the desire to win, the drive to enjoy personal rewards. But I have also voiced caution about the risks of runaway arrogance, ambition, and greed—those ancient failings of unfettered desire that, left unchecked, lead to outrageous violations of decency and trust. As noted, many of us have great hope—for prosperity, possessions, happiness, or just a better way of life. Many of us also like to win—and, yes, it may be just for the sake of winning, but it's hugely important. And many are consumed with an energetic desire to achieve, or perform, whatever is "next." But with each of these natural desires and impulses, there are the equally natural and devilish impulses to cheat, to dissemble, to cut corners, to skirt regulations or break laws.

As the stakes become higher, so too do the temptations. It takes an order of magnitude of courage for a highly ambitious, enormously well-paid executive in a profit-obsessed organization to acknowledge losses or mistakes, much less crimes and misdemeanors. But unless true malefactors are punished—swiftly, definitively, and appropriately—there is a real risk that the guilty and innocent alike will be tarred with the same brush. Thank goodness that Bernie Madoff went to jail. More need to.

2. Enforce Existing Regulations

Whenever a Wall Street crisis occurs, there is a predictable political-posturing response. Congresspeople, editorialists, and pundits rage against the wrongdoers, pledging that steps must be taken so that this sort of awful thing doesn't happen again. What we need, politicians conclude, are more laws and regulations to prevent this badness. Then they can return to their districts with evidence of their responsible leadership.

But what if badness-preventing laws, already in place, are not enforced or are ignored? What good does it do to pile on additional rules and regulations accompanied by equally lax enforcement?

More than once in previous chapters, I have used the analogy of fair rules to help describe what's needed for a level playing field. Whether you're entering the lottery or playing pro baseball, the issue of fairness is part of the game. Without it, what we have is arbitrariness, corruption—the teacher who can be bribed with chocolates to give high grades, the congressperson who can be swayed with huge donations to fiddle with regulations. Fairness requires rules of play that everyone can recognize. But something more: it requires enforcement of those rules. The umpires and referees who maintain the rules and point to infractions must be empowered by the owners, teams, and players to carry out their roles. There must be enough of them on the field to make sure they get the job done. And they must understand the responsibility they owe to the reputation of the sport as well as to those who participate in it.

Somehow, we've short-changed the clout, authority, and efficiency of the referees and umpires who are meant to enforce the rules and regulations of our financial system. I say "somehow," yet what happened is absolutely clear. The regulatory agencies assigned to police the system are woefully understaffed. They have limited ability to intercede to prevent flagrant abuses, much less enforce regulations that are already on the books. As the financial world is rocked by one scandal after another, it has become abundantly clear that people we have employed to maintain the rules of play are only capable of the most cursory oversight. When they finally throw the flag on the field, it's usually long after the play is over.

If I'm mixing my analogies here, it's entirely intentional. We've hired enforcers who can't do their job. Then we blame them, in the aftermath of each scandal, for not doing their job. And, finally, rather than increasing their authority and giving them the tools, resources, and personnel they need to police the streets of the financial world, we cry out for new rules and regulations.

Well, now we have over 800 pages of the Dodd-Frank Act. (And just months after it was passed, MF Global, headed by former New Jersey Governor and former Chairman of Goldman Sachs, John Corzine, failed miserably and lost over a billion dollars of clients' money... doing exactly the things Dodd-Frank was written to prevent.) As each lawyered-up court case has made its way through the justice system, it has become increasingly clear to all of us that the agencies responsible for preventing financial collapse and averting disaster have been fiddling while Rome burned.

For the financial community, there are dire consequences when lax enforcement is combined with the perpetual redesign of regulations. Imagine the climate of uncertainty that hovers over a game where rules and boundaries keep changing, and referees sometimes look the other way. That's the picture. Confusing for the players; baffling to spectators. It's not surprising, then, that trust in our financial institutions has fallen so sharply. These institutions are operating on a playing field that is subject to earthquakes and aftershocks as rules change, boundaries shift, and referees are snoozing and boozing. Enforce the securities laws on the books and maybe we won't need so many new ones.

3. Celebrate Success

America has been the place where stunning success happens, and those who achieve it are lauded. Many are envied, some are revered. And they offer themselves as examples to all that success is possible and that we might achieve it too!

About the time I was wrapping up this book, I was talking to a friend who manages one of the largest investment firms in the country. Aggrieved by the fallout from scandals in the financial markets, he said, "You know what, Farr—*we* didn't do any of those things that these other guys did. We've worked hard for our customers, we haven't broken any laws, we didn't get involved in any of this mess. Yet we're tarred with the same brush."

"Well, whose fault is that?" I responded.

He didn't want to hear it—but I was serious. Here's a prominent executive of one of the leading financial organizations in America, and he's allowing his company to be undeservedly lumped in with the reckless gamblers who crashed the U.S. financial system. This gentleman is successful. So is his company. So are their achievements. Let it be shouted from the rooftops.

America is built on the hard work and determination of creative, aspiring Americans. Generations of us have fought and labored and built and created. We've been focused on providing a *better* life for our children and grandchildren. But the word "success" has been so tainted by those who flaunt their ambition and greed that it's in danger of sounding like poison.

Real success is *not* poisonous. As I'm finishing this book, the summer Olympics are going on. Young, determined, healthy, strong people from around the world are competing against each other, against the clock, and against all previous records. In many cases they show that what had been considered impossible is now possible. We are inspired, and our hearts swell with pride for the magnificent achievements wrought by remarkable talent, hard work, heart, and tenacity.

Feels good—doesn't it? And, of course, we're rooting for the home team, those American kids and young adults who perform feats on the track and field and court and parallel bars that defy belief. If this makes us proud of our homeland, so be it. But there are many other kinds of success to celebrate here. The business start-up that turns into a full-scale, farflung enterprise. The community garden that, each year, yields an abundance of bouquets and crops of homegrown veggies. The college degree attained by the son or daughter of a blue-collar family. The corner restaurant in its twentieth year of operation. The grant for a new wing of the hospital. The buying of a first house or car. The big vacation after many years of hard work. The saving of a landmark building. The opening of a new bike path. The place in the parade for members of the volunteer fire department. It may seem like these success stories can't hold a candle to the winning of an Academy award, or a high place on Forbes most-wealthy-Americans list, or a first place in "American Idol"— but all these success stories are very American and very important.

There is no reason to vilify those who have been successful in earning a great deal of money, or resent those things they've earned by their success. It's okay to dream of a really cool expensive car and a beautiful house and flying first class and drinking champagne. Those who work hardest and smartest, against all odds, should be able to celebrate their success in America.

Investing in our success is the best investment we can make.

4. Remember Our American Dream

America has always been known as the land of dreams and a land where some of those dreams come true. Some of our Founding Fathers' dreams are taken for granted today— freedom to practice your own religion, speak your mind, or print your opinions in the newspaper. Others are unattainable—universal happiness, health, prosperity. But in-between what we have been given and what we cannot possibly attain is the essential glue that holds us together. It's the dream of a land of opportunity—*the hope for what's possible and also for the conditions that will allow you to succeed.*

In 1959 the media had a heyday when Soviet Premier Nikita Khrushchev was escorted through the "kitchen exhibit" in the company of Vice President Richard M. Nixon. With considerable pride, Nixon showed Khrushchev the marvelous conveniences then offered by American manufacturers to the "average" homebuyer— from electric mixers and can openers to dishwashers, washing machines, and dryers. Comparing the average American home to that of the average Soviet citizen, Nixon argued persuasively to convince Khrushchev that capitalist achievements far outstripped those of the socialist regime. Khrushchev, expressing his disdain for the show-and-tell, responded, "This is what America is capable of, and how long has she existed? Three hundred years? One hundred and fifty years of independence and this is her level? We haven't quite reached forty-two years, and in another seven years, we'll be at the level of America, and after that we'll go farther. As we pass you by, we'll wave 'hi' to you, and then if you want, we'll stop and say, 'please come along behind us.' ...If you want to live under capitalism, go ahead, that's your question, an internal matter, it doesn't concern us. We can feel sorry for you, but really, you wouldn't understand."

In debating the relative worth of kitchen appliances, both Nixon and Khrushchev missed the point. What mattered, then and now, was not the number of gadgets and gizmos, but the ability to think about what comes next—what will be possible in the generations ahead—and how to make that a reality. There, the one-upsmanship stops. The American dream begins. Our strategy is hope, but hope as I have described it in this book (with thanks to Professor Snyder)—having the willpower and waypower to get to our destination.

Today, we have no shortage of projects that require a superabundance of hope. We have to figure out how to stabilize our financial system and reduce the national deficit. We have to grow our economy and, at the same time, discover ways to use our resources more efficiently and with vastly reduced damage to the environment. We have to increase educational opportunities, decrease criminality and incarceration, control health-care costs, maintain or reconfigure our infrastructure, and develop more efficient forms of transportation. We have to improve our security and diplomacy, expand employment opportunities, resolve immigration issues, provide better care for the aged, the handicapped, for veterans, for the chronically ill. We have to open new opportunities for entrepreneurism, for inventiveness, for capital investment, for scientific and medical research, for artistic exploration.

And that's not the half of it.

There's lots to do. And no one to tell us exactly how, or when, to do it. Which is the essence of what makes things possible in this country. The possibilities and the opportunities stretch as far as our imaginations can reach.

This dreaming is not only possible, it's necessary. The best investment we can make, today, is in our own ability to solve problems, come up with creative solutions, discover opportunities, and lead the missions that will take us to shared prosperity.

Occasionally, we do have leaders who not only advocate that mission but describe it in words that are worth remembering. I often quote the eloquent John F. Kennedy, who said, "Ask not what your country can do for you—ask what you can do for your country."[7] And Ronald Reagan's words: "You and I have a rendezvous with destiny. We will preserve for our children this, the last best hope of man on earth, or we will sentence them to take the first step into a thousand years of darkness. If we fail, at least let our children and our children's children say of us we justified our brief moment here. We did all that could be done."

Left and right, Democrat and Republican, they dared to talk about a dream that has been shared in this country for more than two hundred years. It remains to be seen whether left and right can work together to repair the broken dikes. But my hopes are high, and from what I've observed of the American spirit, we are within reach of what's possible. Beyond duty, honor and country, we have the conditions that will allow us to succeed: Hope, Trust, Striving, Yearning,

Determination and Grit. This is the stuff of American Dreams. This is the foundation of our Country, our economy and our legacy for future generations. The secret of America's resilience is this indomitable spirit that dwells in American hearts, minds, and souls. It is the gift of providence; it is all we need; and it will see us through to many bright, shining, glorious tomorrows.

Afterword

By P. J. O'Rourke

Restoring Our American Dream is about God. This is a personal opinion that comes as much from knowing Michael Farr as it does from reading what he's written. Michael isn't trying to convert anyone. ROAD isn't a religious tract. I don't mean that Michael intended his book to be about God. I mean that he couldn't help it.

At the risk of being accused of bad economics and worse theology, I claim God is inseparable from the free market. This is not a claim Michael makes. But it's a claim he proves.

The proof lies in the Ten Commandments. God has certain things to tell mankind about human freedoms. Michael Farr has certain things to tell business and political leaders about market freedoms. They are the same things.

"You shall have no other gods before me."

All freedoms come with moral responsibilities. There's something more important than money and power. If you worship money and power, you've got a bad religion. A true believer is willing to sacrifice everything for the sake of what he or she believes in. Would you sell your son to make yourself more money? Would you jail your daughter to give yourself more power over her? And yet, with our deficit spending and the growth of our national debt, we have sold our children's opportunities in life and imprisoned their future hopes — all for the sake of keeping tax dollars in our pockets and political prerogatives in our hands.

"You shall not make for yourself a graven image, or any likeness of anything that is in heaven above, or that is in the earth beneath, or that is in the water under the earth."

And note how many of us have mortgages that are under water. The lesson of this Commandment is not to worship objects. The lesson of ROAD is that no matter how much we'd like a new Mercedes-Benz we don't want to shrug off moral obligations to bow down to car payments.

"You shall not take the name of the Lord your God in vain."

Some people think this Commandment is about saying "Goddamn it." What it really forbids is magic and divination. God doesn't like it when people pretend they can make supernatural things happen with their religion. And Michael Farr doesn't like it when people pretend they can make supernatural things happen with their credit swap derivatives.

"Remember the Sabbath day, to keep it holy."

Goddamn it, quit checking your emails at the family dinner table. See the first Commandment, above. When you spend all your time and effort "making it," you aren't paying attention to what kind of it you're making. We get so busy we forget what we're doing. God, in his infinite wisdom, left all of creation on His desk until Monday morning. This is why He made the world while we made a mess.

"Honor your father and your mother."

Unfunded Social Security, Medicare, and pension schemes are hardly the best ways to do this. Material greed and political intransigence cause us to cheat our parents as well as our children.

"You shall not kill."

"You shall not commit adultery."

"You shall not steal."

"You shall not bear false witness against your neighbor."

Imagine a society where lying, thieving, employing brute force to achieve your ends, and betraying the people to whom you have sworn to be true, are considered okey-dokey, or even admired as signs of strength.

"Is this the way to conduct life?" asks God. "Is this the way to conduct free markets?" asks Michael Farr. Now consider Wall Street. Consider Washington.

"You shall not covet your neighbor's house; you shall not covet your neighbor's wife, or his manservant, or his maidservant, or his ox, or his ass, or anything that is your neighbor's."

This may be the most underrated Commandment in the Bible.

And it may be the most important Commandment in Restoring Our American Dream.

God is saying that if you want a McMansion, if you want a cleaning lady, if you want a pot roast, don't envy what the people across the street have – go get your own.

This is God's fundamental statement of free market principles. And, on the subject of these principles, Michael Farr delivers a biblical commentary.

What happens to America when business and politics pervert the free market so that people who want better homes, better circumstances, and better lives don't have a chance to get them? What happens when a large percentage of our citizens are in a situation where they can't avoid violating the Tenth Commandment? What kind of people do they become? What kind of people do we become?

We can listen to the message of *Restoring Our American Dream*, or we can go to hell.

End Notes

Chapter 1

16 "chances for survival are much higher": John Call, PhD, JD, "Developing the Survival Attitude" in Pysychology Today, July 19, 2008.
21 "something special just for me!": http://lyricsplayground.com/alpha/songs/t/thewellsfargowagon.shtml.

Chapter 3

49 "is a powerful inducement to action": from private correspondence with Timothy Keith-Lucas, PhD, Professor of Psychology, Emeritus, Sewanee: The University of the South.
51 "building the enabling conditions of a life worth living": Martin E.P.Seligman, Flourish: A Visionary New Understanding of Happiness and Well-being, Free Press, 2011, Kindle edition, location 179.
51 "what we choose for its own sake": Ibid., location 304.
53 "totally lacking a basis in reality": C.R. Snyder, The Psychology of Hope: You Can Get There from Here, The Free Press, 1994, Kindle edition, location 120.
49 "provides the key to understanding hope": Ibid., location 138.
53 "between an impossibility and a sure thing": Ibid., location 181.
53 "sustaining movement toward desired goals": Ibid., location 205.
54 "the willpower and the waypower to get to our destination": Ibid., location 249.

Chapter 4

70 "a chemical called oxytocin": Paul J. Zak, The Moral Molecule: The Source of Love and Prosperity, Dutton, 2012, Kindle edition, location 89.
74 "difference between the per capita income of the United States and Somalia": Tim Harford, "The Economics of Trust", July 21, 20120, http://www.forbes.com/2006/09/22/trust-economy-markets-tech_cx_th_06trust_0925harford.html .

74 "According to a BBB/Gallup poll conducted in April 2008": "BBB Gallup Trust in Business Index", http://www.bbb.org/us/storage/0/Shared%20Documents/Survey%20II%20-%20BBB%20Gallup%20-%20Executive%20Summary%20-%2025%20Aug%2008..pdf.

Chapter 5

76 "some money back to your unseen benefactor": Paul J. Zak, The Moral Molecule: The Source of Love and Prosperity, Dutton, 2012, Kindle edition, location 322.

77 "exactly where that advantage lies": Ibid., location 304.

80 "no matter how your 'partner' behaves in the future": "Promises, Lies and Apologies: Is It Possible to Restore Trust," July 26, 2006, Knowledge@Whartonm http://knowledge.wharton.upenn.edu/article.cfm?articleid=1532.

85 "the level of trust inherent in the society": Francis Fukuyama, Trust: The Social Virtues and the Creation of Prosperity, Simon & Schuster, 1995, Kindle edition, location 188.

86 "that trust contributes to economic, political and social success": Michael Kosfeld, Markus Heinrichs, Paul J. Zak, Urs Fischbacher, Ernst Fehr, "Oxytocin increases trust in humans," Nature, Vol. 435, No. 2, June 2005, pp. 673-676.

86 "Trust in banks fell 24 percent": http://www.gallup.com/poll/155357/Americans-Confidence-Banks-Falls-Record-Low.aspx.

86 "Trust in the media plummeted 14 percent": http://www.gallup.com/poll/157589/distrust-media-hits-new-high.aspx.

86 "confidence in U.S. public schools declined nearly 10 percent": http://www.gallup.com/poll/155258/Confidence-Public-Schools-New-Low.aspx.

86 "from around 30 percent to 11 percent over a similar period": http://www.gallup.com/poll/141512/congress-ranks-last-confidence-institutions.aspx.

90 "36 percent of Americans did not vote": http://www.census.gov/prod/2010pubs/p20-562.pdf.

90 "having direct impact on local communities": Bridget Hunter, "2011 U.S. State, Local Elections Important Despite Low Turnout," IIP Digital, November 9, 2011, http://iipdigital.usembassy.gov/st/english/article/2011/11/20111109172024tegdirb0.4291651.html#ixzz1x6c9PQYbhttp://iipdigital.usembassy.gov/st/english/article/2011/11/20111109172024tegdirb0.4291651.html#ixzz1x6c9PQYb.

Chapter 6

94 "Petitioned for Redress in the most humble terms": http://www.ushistory.org/declaration/document/index.htm.
94 "to petition the Government for a redress of grievances": http://constitutionus.com/.
98 "that lobbying and corruption are substitutes": Bård Harstad, Jakob Svensson, "Bribes, Lobbying, and Development", American Political Science Review, Volume 105, Issue 1, February 2011.
98 "high-level bribery, lobbying or influence peddling": Daniel Kaufman, "Corruption and the Global Financial Crisis," Forbes.com, 01/27/09, http://www.forbes.com/2009/01/27/corruption-financial-crisis-business-corruption09_0127corruption.html.
100 "balance between Republican and Democratic parties and candidates": Lindsay REnick Mayer, Michael Beckel, Dave Levinthal, "Crossing Wall Street," November 16, 2009, http://www.opensecrets.org/news/2009/11/crossing-wall-street-1.html.
100 "among the top ten donors to national and state politics": OpenSecrets.org, "National Donor Profiles," http://www.opensecrets.org/orgs/index_stfed.php.
100 "the top spenders among lobbying clients": OpenSecrets.org: Center for Responsive Politics, http://www.opensecrets.org/lobby/top.php?indexType=s.
100 "this entire sector spent $140 million": OpenSecrets.org: Center for Responsive Politics, http://www.opensecrets.org/lobby/indus.php?id=Q&year=2011.
100 "well over half a billion dollars": OpenSecrets.org: Center for Responsive Politics, http://www.opensecrets.org/lobby/top.php?showYear=2011&indexType=c.

101 "to the Wal-Mart that meets the private-needs requirements of foreign government officials": David Barstow, New York Times Business Day, "Vast Mexico Bribery Case Hushed Up by Wal-Mart After Top-Level Struggle," http://www.nytimes.com/2012/04/22/business/at-wal-mart-in-mexico-a-bribe-inquiry-silenced.html?pagewanted=all

Chapter 7

105 "to choose one's own way": Viktor E. Frankl, Man's Search for Meaning, Beacon Press, 2006.
105 "garage of his home and hanged himself": New York Post, June 19, 2012, pages 18-19.
106 "surviving one of America's deepest recessions": Brianna Karp, The Girl's Guide to Homelessness, Harlequin, 2011.
110 "and the value of that gain to us": Dan Gilbert, "Why We Make Bad Decisions," TED Conference filmed July 2005 posted December 2008, http://www.ted.com/talks/dan_gilbert_researches_happiness.html.
115 "overestimated the value of our present pleasures": Dan Gilbert, Ibid.

Chapter 8

121 "As reported by Matt Taibbi": Matt Taibbi, "The Scam Wall Street Learned from the Mafia," Rolling Stone, June 21, 2012, http://www.rollingstone.com/politics/news/the-scam-wall-street-learned-from-the-mafia-20120620.
122 "income from its investment as a result of the lower interest rate": Ibid.
122 "for defrauding cities and the U.S. Internal Revenue Service": Bob Van Voris, "Ex-GE Bankers Convicted of Municipal Bond Bid-Rig Scheme,"Bloomberg Businessweek, October 18, 2012, http://www.businessweek.com/news/2012-05-11/ex-ge-bankers-convicted-of-municipal-bond-bid-rig-scheme.
125 "who cannot help but be dysfunctional and corrupt": Christopher Hayes, Twilight of the Elites: America After Meritocracy, Crown, 2012, Kindle edition, location 260.

126 "had any influence at all on roll-call votes": Ibid., location 2108.

126 "share decisions having at least national consequences": C. Wright Mills, The Power Elite, Oxford University Press, 2000.

126 "in the literal sense, to their fellow meritocrats": Christopher Hayes, Twilight of the Elites: America After Meritocracy, Crown, 2012, Kindle edition, location 2158.

127 "produces a worse caliber of elites": Ibid., location 2248.

127 "within view and yet remains out of reach": Ibid., location 2257.

Chapter 9

135 "confidence in banks hovered around 23 percent": Gallup Politics, http://www.gallup.com/poll/155258/Confidence-Public-Schools-New-Low.aspx.

137 "and 12 percent trusted Congress": http://www.gallup.com/poll/155258/Confidence-Public-Schools-New-Low.aspx.

140 "newly conceived and constitutionally created central government": Ron Chernow, Alexander Hamilton, Penguin Books, 2004, Kindle edition, location 6634.

141 "values of the varied bills and coins in circulation": Ibid., location 4654.

141 "measured by percentage of GDP": "First Report on the Public Credit" in Wikipedia, http://en.wikipedia.org/wiki/First_Report_on_the_Public_Credit.

141 "attractive fixed rate of interest": Ron Chernow, Alexander Hamilton, Penguin Books, 2004, Kindle edition, location 6814.

141 "retired about 5 percent of its debt every year": Ibid., location 6764.

142 "affected by appearances as well as realities": "Report on Public Credit, vol. 6, p. 96, January 1790, quoted in Chernow, location 6764.

142 "accompanied with the means of extinguishment": Chernow, location 6829.

142 "first announcement of the new policies": Chernow, location 6847.

142 "speed never before seen in America": Chernow, location 6895.

143 "the whole system of public credit": Chenow, location 8128.

143 "reasonable dimensions or it will tumble": Alexander Hamilton, letter to William Seton, February 10, 1792, quoted in Chernow, location 8610.

143 "restored temporary calm to the marketplace": Chernow, location 8625.

143 "rules that would govern securities trading": Chernow, location 8691.

144 "to protect the credit rights of customers.": "Federal Reserve System" in Wikipedia, http://en.wikipedia.org/wiki/Federal_Reserve_System.

146 "with advantage to themselves and to the community": Alexander Hamilton, "Report on the Bank" quoted in Chernow, location 7892.

147 "People were human beings to him": Philip VanDoren Stern, script for It's a Wonderful Life (1946).

Chapter 10

149 "you'll end up someplace else": http://www.goodreads.com/author/quotes/79014.Yogi_Berra.

151 "their hooves while their mandibles manage grass control": "Flood Control in the Netherlands" in Wikipedia, http://en.wikipedia.org/wiki/Flood_control_in_the_Netherlands.

152 "their own elections and levy their own taxes": Ibid.

152 "blessings of liberty to ourselves and our Posterity": http://constitutionus.com/.

155 "contribute to the debate that dictates policy": Douglas A. Kass, commencement address, Alfred University, May 15, 2010.

155 "U.S. society is generally fair and decent": Rasmussen Reports, March 20, 2012, http://www.rasmussenreports.com/public_content/politics/general_politics/march_2012/64_say_u_s_society_is_fair_and_decent.

155 "an enduring document that remains relevant today": The AP-National Constitution Center Poll, August, 2011, Conducted by GfK Roper Public Affairs & Corporate Communications.

156 "seem deeply offensive to most people": Ibid.

156 "solve our problems and get what we want": http://www.people press.org/2012/06/04/partisan-polarization-surges-in-bush-obama-years/.

156 "during decades of rising Gross National Product (GNP) per capita": "World Happiness Report" edited by John Helliwell, Richard Layard and Jeffrey Sachs, http://www.earth.columbia.edu/sitefiles/file/Sachs%20Writing/2012/World%20Happiness%20Report.pdf.

162 "We can feel sorry for you, but really, you wouldn't understand": "The Kitchen Debate" (1959) on TeachingAmericanHistory.org, http://teachingamericanhistory.org/library/index.asp?document=176.

163 "ask what you can do for your country": http://www.famousquotes.me.uk/speeches/John_F_Kennedy/5.htm.

164 "We did all that could be done": http://www.fordham.edu/halsall/mod/1964reagan1.html.

Acknowledgments

I am very grateful to so many friends and colleagues who helped with this book. Laurie, Robert and Maggie, thank you for helping me through.

My great thanks to Ed Claflin. Ed is a true friend, superb writer, and great American. I never want to undertake another project without Ed. Thanks to Scott Sobel who always encourages and prods me to write. Thank you to my publisher Cathy Teets at Headline Books and to my dear friend, the wonderfully talented author Karna Small Bodman for her constant support and for introducing me to Cathy.

When I spoke with General Brent Scowcroft about writing a Foreword, he said no. He said he'd like to hear about the book but he simply doesn't write Forewords. After he heard about our focus on restoring trust and hope to America, he said he'd write. He said he thought that the book had a very important message. I'm humbled and enormously grateful to this most accomplished American hero. Brent Scowcroft has lived "Duty. Honor. Country." more completely than anyone I can think of.

P.J. O'Rourke is as good a man and as faithful a friend as any man could want. His counsel and support have carried me through book number three. Thank you, P.J. for always being there, saying yes, and for writing a delightful Afterword.

Dr. Timothy Keith-Lucas, Ph.D. from The University of the South in Sewanee, Tennessee provided outstanding wisdom, expertise and insight about behavioral psychology.

Thank you to Nik Deogin and all of my friends at CNBC for your encouragement. John D. Haesler from the University of Michigan was a fabulous research analyst and strategist. Thank you to Igor Dabik, Corky Crovato, Ed Eckenhoff, and Hector Garcia for sharing your American Dream stories (see www.ouramericandream.com).

Finally, thank you to everyone at Farr, Miller & Washington: Susan Cantus, Sunny Miller, John Washington, Taylor McGowan, Caroline Savage, Keith Davis, Michael Fox, Glenn Ryhanych, Darshan Gulati, Jennifer Gregory, Andy Mathes, Buck Montgomery, Steven Stone, Javier Madariaga, Katherine Kawecki, Grace Santos, Sheldon Cohen, Chris White, and Joe Coreth.

About the Author

Michael K. Farr is President and majority owner of Farr, Miller & Washington, LLC. He is Chairman of the Investment Committee and is responsible for overseeing the day to day activities of the firm. Prior to starting FM&W, he was a Principal with Alex Brown & Sons.

Mr. Farr is a paid Contributor for CNBC television and has appeared on *The Today Show, Good Morning America, NBC's Nightly News, CNN, Bloomberg, Reuters,* and the *Nightly Business Report.* Mr. Farr is heard on Associated Press Radio and National Public Radio, and he has been quoted in the *Wall Street Journal, Forbes, Fortune, The Washington Post, Businessweek, USA Today,* and many other publications. He is a member of the Economic Club of Washington, DC, National Associaton for Business Economics, The World Presidents' Organization, and The Washington Association of Money Managers. He is the author of *A Million Is Not Enough*, published by Hachette Book Group USA in 2008, and *The Arrogance Cycle,* September 2011 by Globe Pequot Press.

Mr. Farr is the Chairman of the Sibley Memorial Hospital Foundation. He also serves on the Board of Trustees at Sibley Hospital; he is the former Vice Chairman of the Board of the Salvation Army; he is a former member of the Board of Trustees of Ford's Theatre; he is the former Chairman of the Board of Directors of the Traveler's Aid Society, Nation's Capitol Progress Foundation, and the Paul Berry Academic Scholarship Foundation, as well as a member of the Board of the Neediest Kids. Mr. Farr is a graduate of the University of the South in Sewanee, Tennessee. He is married and has two children.